Preparing and Delivering Scientific Presentations

John Giba · Ramón Ribes

John Giba · Ramón Ribes

Preparing and Delivering Scientific Presentations

A Complete Guide for International Medical Scientists

Springer

John Giba
C/ Cau Palau, 54
08181 Sentmenat
Spain
john_giba@yahoo.com

Ramón Ribes
Platero Martinez, 19
14012 Córdoba
Spain
ribesb@gmail.com

ISBN 978-3-642-15888-9 e-ISBN 978-3-642-15889-6
DOI 10.1007/978-3-642-15889-6
Springer Heidelberg Dordrecht London New York

Cover design: eStudioCalamar Figueres/Berlin

Printed on acid-free paper

Springer is part of Springer Science+Business Media (www.springer.com)

"To my father, John Giba, who showed me there was nothing to fear. To my wife, Núria Fernàndez Bayó, and to my daughter, Júlia Giba Fernàndez, who show me there is everything to hope for."

John Giba

"To my late father, Ramon Ribes Blanquer, full professor of Anatomy, to whom I owe many of the ideas inspiring this book"

Ramón Ribes

Preface

After the publication of *Medical English* (R. Ribes, P.R. Ros: Springer. Nov 2005) and *Radiological English* (R. Ribes, P.R. Ros: Springer. Nov 2006), Ramón Ribes became the first editor and coordinator of a series of books on medical English for different specialties. These books are mainly written by non-native-English-speaking doctors who know how hard it can be to learn a second language and understand the challenge of dealing with the acronyms, abbreviations, and jargon that are so important in medical English.

Learning scientific and medical English is extremely demanding. A sound knowledge of English grammar lays the foundations on which you can build your medical English. Just as physicians of all specialties need to know anatomy and physiology (i.e. the normal structures of the body, how they relate to one another, and their functions), professionals from non-English-speaking countries who want to learn English will need to become familiar with different linguistic structures and their functions. Furthermore, Latin and Greek terminology are so prominent in medical English that some basic knowledge of these "dead" languages is essential to speak and write medical English properly. Being fluent in medical English implies being able to communicate effectively in speech and in writing in a wide variety of situations with a wide variety of interlocutors. This book focuses on one small but important aspect of medical English: formal scientific presentations.

Nobody doubts that English is the language of science and medicine, and nowhere is this more evident than at international meetings and courses. If a researcher or physician wants to communicate his or her results and conclusions to the scientific community, he or she must do so in English. Many careers have been stunted by poor English, and many more have never gotten off the ground because physicians failed to take advantage of the opportunity to speak at a meeting because of fears that their English was not up to the task. Indeed, presenting an oral communication at an international congress can be daunting for even the most fluent of non-native-English-speaking doctors. Yet, at the same time, with proper preparation, even relatively weak speakers can rise to the challenge.

We have watched and listened to over one thousand scientific presentations at American and European courses and congresses. We have also had the experience of speaking to a wide variety of audiences on many occasions. We would like to share our experience both as members of the audience and as lecturers, and we sincerely hope this humble manual helps you improve the overall quality of your presentations.

John Giba
Ramón Ribes

Acknowledgment

This book would not have been possible without the help and cooperation of many people. Special thanks to Núria Fernàndez for elaborating the diagrams and drawings in the example slides and to Xavi Calvet, MD, PhD for his critical review of the manuscript and helpful suggestions.

Contents

Chapter 9
Conversation Survival Guide

Chapter 10
Improving Your Pronunciation

Chapter 11 Appendices

CHAPTER 1

Chapter 1 To Speak or Not to Speak? Questions You Should Ask Yourself before Agreeing to Do a Talk in English

The most important thing in any scientific communication, whether oral or written, is its contents. A maxim usually attributed to a Spanish proverb advises "Don't talk unless you can improve on the silence". Indeed, if you don't have anything important to say, it's better to say nothing at all. However, if you have presented an abstract to an international congress that has been selected for an oral communication or if you have been invited to speak at an event in English, the odds are that you have some relevant ideas that will interest your colleagues.

A physician or scientist will have no difficulties deciding whether to do a talk in his or her own language. Every time you decide whether or not to do a given talk in your own language, you consciously or unconsciously consider many questions about the situation. In deciding whether to give a talk in English, you will have to consider these very same questions; however, given the difficulties involved in presenting in a language other than your own, you would be wise to consider all these aspects consciously and conscientiously.

However, there is one crucial question that you will undoubtedly ask yourself about giving a talk in English that does not come into play when you need to decide whether to give a talk in your own language: Can I communicate my ideas effectively enough in English? The key words in this question are "effectively enough". Few non-native English speakers can hope to communicate in English with the ease and skill that they can in their own language. Whereas you may be capable of being spontaneous, entertaining, and witty in your own language, you may have to content yourself with being clear, coherent, and informative in English. This is no small feat. Many will doubt their ability to achieve even this rudimentary level of success, but if you are one of these you should think again.

Even if your general knowledge of English leaves much to be desired, it is important to remember that giving a successful talk in English may not require general proficiency in the language. Opera singers often sing in languages in which they have no idea about grammar or even the meaning of individual words! Although we do not think that it is a very good idea to attempt a presentation in English if you do not have any knowledge of the language whatsoever, many presenters with weak English have been able to give strong presentations.

Indeed, your scientific and medical English is probably better than you think. While many physicians would have difficulties talking about everyday activities or their families, most would hold their own much better in a conversation about

J. Giba and R. Ribes, *Preparing and Delivering Scientific Presentations*,
DOI: 10.1007/978-3-642-15889-6_1, © Springer-Verlag Berlin Heidelberg 2011

their specialty and would fare surprisingly well in their areas of special interest. In all probability, you will already be familiar with most of the English vocabulary related to the topic you are going to speak about – to acquire the knowledge necessary to add something to the common body of scientific knowledge you will have read many articles in English.

Over 2000 years ago, Cato the Elder gave this advice "Grasp the subject; the words will follow." Although it always pays to work at improving your English, you are unlikely to have enough time to make much headway in your general English or even in your general medical and scientific English between the time when you agree to give a talk and the talk itself. However, you can make huge improvements in the language you need for your presentation. Furthermore, preparing your presentation will help you make enormous gains in your scientific and general English. So, surprisingly, while working hard at improving your English may not immediately help you improve your presentation, working hard at improving your presentation will undoubtedly help you improve your English!

Finally, bear in mind that although it is extremely difficult to speak a foreign language without mistakes, nobody will expect you to speak perfect English and nobody will judge you for minor mistakes that do not affect the audience's comprehension. There is a long tradition of heavily accented speech by highly respected foreign scientists in English-speaking countries, so having an accent or making a few minor pronunciation or grammatical mistakes might even enhance your prestige – as long as you have something interesting to say.

Below follows a list of questions that you should ask yourself before agreeing to do a talk in English.

Whenever you are invited to present a paper, there are at least ten questions you have to ask to yourself:

1. What is at stake? What do I stand to lose or gain from accepting this challenge?
2. Do I have something interesting to say about this topic?
3. Who will I be speaking to?
4. Will I be happy talking about this topic to this audience?
5. What type of paper am I presenting?
6. Does my presentation fit in with the type of paper I have been invited to present?
7. Am I able to tell my story in an engaging way?
8. Where will the presentation take place?
9. Are there ulterior reasons why have I been invited to present this paper?
10. Will I be able to find the time to prepare and rehearse this presentation properly?

Let's have a brief look at each of these questions:

1. What is at stake? What do I stand to lose or gain from accepting this challenge?

You don't need a fantastic imagination to come up with several reasons why you shouldn't bother to do a particular talk in English. Just thinking about the diligence and sheer amount of work involved in preparing a presentation and all the things that can go wrong is enough to make any reasonable person want to flat out refuse to do a talk. However, if everybody adopted this attitude, the human race would surely die out as nobody would bother to have children. Indeed, if you approach life with such a negative attitude, you will never accomplish anything.

Think instead of all benefits you can reap from your efforts. Sure, preparing a well-constructed talk involves a lot of work, but this work will help consolidate your knowledge in the area of interest and lay the groundwork for future projects, articles, book chapters, etc. Moreover, preparing the talk can substantially improve both your general and medical English, and much of the preparation that goes into improving your delivery will be transferable to future talks.

A successful presentation at an international conference can help you to make useful contacts, establish your reputation in the field, and jumpstart your career at home as well as open the door to potentially different paths in the future. An outstanding presentation will likely lead to future invitations, and someday you may be the one chairing the session or presiding over a committee. So, perhaps it might help to re-phrase our question as "Can I afford to decline this invitation?"

On the other hand, a poor presentation will reflect badly on you and possibly on your institution and country as well. So, unless you are willing to commit to doing whatever it takes to ensure that your presentation is at least fairly good, you should politely decline the invitation. However, this decision should not be taken lightly – remember that declining an invitation is closing a door that may not be easy to open in the future.

2. Do I have something interesting to say about this topic?

Everything hinges on this question. No matter how good your English presentation skills are and no matter how entertaining your presentation is, nobody will be truly impressed if you have nothing new and interesting to convey. On the other hand, audiences are willing to forgive many shortcomings in presentation (including a strong accent, lack of fluency, and even mispronunciation of key words) provided the content of the talk is truly interesting.

3. Who will I be speaking to?

Although this question probably shouldn't influence your decision about whether to agree to give a presentation, it is perhaps the most important question to bear in mind when you are preparing your presentation. The level of the talk will be different if the audience is made up of residents than if it is made up of senior specialists. Heterogeneous audiences are especially challenging, as you need to make your talk interesting both for those who have in-depth knowledge about the topic and for those who only have a vague idea about it. Speaking at the wrong depth is one of the main reasons presenters fail to connect with the audience. It is essential to know as much as possible about the audience before preparing your presentation, so be sure to ask the chairman this important question before accepting his invitation to participate in the meeting.

4. Will I be happy talking about this topic to this audience?

If you have presented an abstract to an international congress, we hope that it will deal with a subject you feel comfortable talking about. On rare occasions, you may have submitted an abstract for a poster presentation that the organizers found so interesting that they earmarked it for an oral presentation. In any case, you are to be congratulated if the organization considers your abstract worthy of an oral presentation.

If the organization has invited you to speak, you will probably have already achieved a certain level of recognition in your field, either from previous talks and/or articles or because somebody on the committee knows you or somebody who knows your work. However, most invitations to talk come with some prescriptions. Often you are assigned a topic and a timeslot (for example, a 40 min talk about large-vessel vasculitides) and have no choice whatsoever in the matter. Other times you are given a general area from which you can choose a specific topic. You will rarely be offered the opportunity to choose your talk freely.

If you decide to decline an invitation to give a presentation because you are uncomfortable with the topic, do so immediately to enable the organizers to look for somebody else to replace you. Be sure to explain why you are declining the invitation and let them know if there is a related topic you would be willing and able to contribute on and that you would be pleased to participate in future meetings provided you consider yourself able to make a valuable contribution. The organizers will appreciate your frankness and may even be willing to change the specifications for your talk.

5. What type of paper am I supposed to present?

Talks at international congresses can last from five minutes to an hour. The amount of preparation and rehearsal will vary accordingly. Likewise, the stakes – the potential benefits of a good talk, as well as the potential for disaster if you are inadequately prepared – increase with increased exposure.

Will you be expected to field questions from the audience after your talk? If so, to what extent can you predict the questions that the audience might ask? Many physicians would be willing to give a presentation in English if they could be assured that they would not be required to deal with questions from the audience afterward. We understand these concerns and Chap. 6 “The Dreaded Questions and Comments Section” will help you deal with this issue.

Will there be a questions and comments section in which you are expected to interact with the rest of the panelists afterwards? Will you be expected to discuss the topic with other experts or even to debate them? This situation obviously requires much more developed English language skills, as it is much more difficult to predict what you will need to say.

The most common types of scientific presentations are described in more detail in Chap. 2.

6. Does my presentation fit in with the type of paper I have been invited to present?

If you have already prepared and delivered a particular presentation in English or even in your own language, you need to consider whether it fits in with the type of paper you have been asked to present. On the other hand, when you know a topic well (and we assume that you do know it very well if you have already prepared and presented it), you can always redesign your presentation to fit different formats. However, this can require a considerable amount of work and you must be prepared to take on the challenge before you accept. And don't forget – it is as least as difficult to transform a 40-minute talk into a 7-minute presentation as it is to expand a 7 min presentation into a 40-minute talk. More importantly, whether or not you have to change the length of your talk, you will probably have to change the depth of your talk to suit your audience.

Even if you are invited to give the exact same talk that you have given before to a similar audience, it is very important to rehearse if it has been some time since you last gave it. You need to try to make it fresh and new, both for your own sake and in case there is anyone in the audience who has heard it before. Be sure that your material is up to date – things change quickly in medicine and information that was valid a couple of years ago may be obsolete today.

7. Am I able to tell my story in an engaging way?

While not everybody can enthrall the audience every time they rise to the podium, you must have a sufficient interest in your topic to make it come alive for others. If you fail to engage the audience, you run the risk of wasting their time and ultimately of wasting your own as well. Again, it is important to remember that you are not being asked to entertain the audience and that what you have to say is more than important than how you say it

8. Where will the presentation take place?

Venues can be quite important. Traveling long distances to speak can be tiring and expensive; however, if you have submitted an abstract for a congress, you are well aware of any inconveniences involved in traveling. On the other hand, traveling to congresses is an opportunity to visit new cities, and even the busiest of agendas will usually allow you some time to get to know something of the host city. Furthermore, delivering a presentation at certain meetings will give you added confidence when you present papers elsewhere in the future. Giving a talk at a World Congress, an American Congress, or a European Congress of your specialty will enable you to give talks anywhere.

9. Are there ulterior reasons why I have been invited to present this paper?

Do I really deserve the invitation? Have I been invited for "regional" reasons? You may be invited to give talks for a myriad of reasons. Ramon was invited to one of his first talks in English for "regional" reasons. The American College of Physicians of a certain state in the United States organized a course in Spain in which 20% of the lecturers had to be Spanish radiologists. After most of his colleagues declined the invitation, he presented two papers, because although he had just arrived onto the radiological scene, his fluency in English allowed him to deliver the presentations. We honestly think that the reason for which you have been invited to present a paper is not that important: what really matters is that you feel confident enough to deliver it properly

10. Will I be able to find the time to prepare and rehearse this presentation properly?

We have saved this crucial question for last. Preparing a scientific presentation takes time, and preparing and rehearsing a presentation in a language other than your own takes a lot of time. Whether you are to give a five-minute presentation or teach a one-hour class, you will need to devote many hours to preparing for the brief time when you are in front of the audience. Even if you already have a similar presentation in your own language or even in English, in all probability you will need to adapt it to fit the audience and context as well as to rehearse the new version. Your success or failure depends on your willingness to commit to proper preparation, so be sure to consider this question carefully.

CHAPTER 2

Chapter 2 Preparing a Talk

Introduction

The importance of proper preparation cannot be exaggerated. Whether you have agreed to speak for five minutes or five hours, you can be sure that the time you actually spend delivering your presentation to your audience will only be a very small fraction of the time that goes into your talk. It takes a lot of time and effort to create a good presentation, and you will probably discover that your ideas will evolve as you work on your talk. Because the most important aspect of any scientific communication is the ideas it aims to convey, the most important tasks are deciding what to include (and what to leave out) and how to organize the information you want to communicate. Next in importance comes the preparation of the slides that will serve as the guide to your talk – both for you and for your audience. Finally, you can improve your delivery by rehearsing assiduously. Just as your ideas will evolve as you work on your talk, working on your delivery will allow you to discover the strengths and weaknesses of your talk. Thorough preparation is the key to success.

This chapter will give you a few tips on researching, organizing your ideas, creating your slides, and rehearsing for your presentation.

Before You Start

There are a couple issues that you should consider as soon as possible after agreeing to do a talk. You should adjust the style and length of your talk to fit both the type of presentation you are expected to give and the time slot when you are expected to give it.

Types of Presentations

Broadly speaking, whenever you speak in front of an audience you are "presenting". In this sense, a scientific presentation could range from something as short and simple as asking a question from the audience at a conference to forming

J. Giba and R. Ribes, *Preparing and Delivering Scientific Presentations*,
DOI: 10.1007/978-3-642-15889-6_2, © Springer-Verlag Berlin Heidelberg 2011

part of an expert panel discussing controversial issues in a particular field. Here we briefly discuss the most common kinds of presentations.

1. *Questions to the lecturer from the audience (mini-presentations)*

Asking a question from the audience at an international congress is a great way to make your debut in the international scientific forum. This experience can help you get over your shyness and prepare you for the day when you will step behind the podium. The question and comments session after a presentation provides you with the opportunity to give a "mini-presentation" without the pressure of having an official role as a speaker.

Although native English speakers can formulate questions off the top of their heads, understanding a lecture and asking a question can be a great challenge for those who are less proficient in the language. Thus it is important to prepare thoroughly if you are going to ask a question. Choose a topic you know well, and read up on it prior to entering the lecture hall. Take notes during the presentation. Design your question in advance. Try to ask a good question on the topic of the lecture with a short introduction so that you are exposed to the audience a bit longer than usual. Do not ask a very short question.

Bear in mind that the very purpose of your question is not the question itself: although neither the lecturer nor the audience will notice it, you are simply gaining experience speaking English in public and learning to overcome stage-fright. Rehearse your question mentally before raising your hand and asking for the microphone.

"Mini-presentations" have a certain protocol. First, you should introduce yourself concisely, stating your name, your hospital, and your country. Then, you should thank the lecturer and praise his or her presentation. Then ask your question. Do not make your point on the topic of the presentation.

A few practical tips:

Being heard. Make sure that the microphone is not so close to your mouth that the audience can hear you breathing and that it is not so far away that the audience cannot hear you at all.

If the microphone is on a base, do not touch it. If you are given a wireless microphone, do not forget to grab it with both hands because trembling – not only of your hands but of your voice – is quite likely in your very first question in public.

Speak up: many beginners speak so softly that neither the lecturer nor the audience can hear them. If your question cannot be heard, it cannot be understood. The panel will ask you to repeat your question and this could make you even more nervous. So, speak a little louder than usual to ensure that everybody can hear you and understand you.

Debating the lecturer. One of the main reasons most non-native English speaking members of the audience do not ask questions is the fear of entering into a debate with the lecturer. However, this is an unfounded fear. Unless you are a prominent figure in the scientific community, it is unlikely that experts will be interested in your opinion about the topic of your question. They will simply answer and proceed to the next question. Remember that, as a general rule, scientific sessions are always behind schedule. Therefore, it is less likely that you will have to respond to the lecturer's answer to your question if you ask the last question in a series of queries than if you ask the first one, simply because the session will be running out of time. See Chap. 3 for an example of a mini-presentation and useful language in this context.

2. *Scientific reports session*

At most conferences, there are several "scientific sessions" in which six or more papers related to a particular area of research are presented in 60 to 90 minutes. In most cases, each speaker presents an overview of a single study in a short talk (the time limit for such presentations ranges from 6 to 10 minutes with an additional 3 to 5 minutes for questions). These sessions are presided over by one or more chairpersons, who introduce speakers and their topics, and oversee the question and answer period. In this type of presentation, it is crucial to stay focused on your particular topic and exclude any information that is not absolutely essential for listeners to get the gist of your study. Most beginners fail by wanting to explain too much – the format does not allow for extensive background information, and the introduction must be very brief. Most general comments can be omitted altogether – it is common (and boring) for a well-versed audience to hear six speakers say something like "breast cancer is the second most common cancer in women". In this kind of talk, it is important to get to the results section quickly. Unless the topic of your paper itself is an innovative new approach to investigating a particular topic, it is the results and their implications that you and your audience will be most interested in. Be sure to leave ample time to go over your conclusions thoroughly.

3. *Symposium*

Like scientific reports sessions, symposia comprise a series of oral presentations with related content. In this case, however, there tends to be a more extensive examination of a single topic than in scientific paper sessions. A symposium will often include fewer (and longer) talks, and speakers' topics will be assigned by the organizers. Ideally, the speakers will have communicated with each other before preparing their talks to ensure that they fit together well and do not overlap excessively. The chairpersons' comments should help to integrate talks, and a discussion among the speakers might be scheduled for after the presentation.

4. *Invited address*

Invited addresses tend to be longer, typically lasting 20 to 60 minutes. In these talks, the speakers might discuss a specific area of research, summarizing and integrating information from several studies. Apart from describing what is known about the topic and discussing research that is currently underway, the speaker might also go into future directions for research.

One special kind of invited address is the *refresher course*. These are intended to review basic concepts about a topic as well as to bring the audience up to date about new developments in the field. In this kind of talk, it is important to remember that many of the attendees will have little or no knowledge of your talk, so you will have to include a thorough introduction.

Slots

Another important aspect to consider before designing your presentation is the time slot you have been assigned, in other words, when you are scheduled to give your talk. Unfortunately, this information is not always available early enough for you to take full advantage of it.

If possible, it can very helpful to obtain a copy of the program for the event you will be participating in to find out the answers to the following questions:

- When are you talking? Which is your slot?
- Who is presenting a paper before you?
- Who is presenting a paper after you?
- Who is who and who are you in the context of the course?

Certain time slots entail special circumstances that must be taken into account. Knowing about them will help keep them from undermining your confidence.

If you are responsible for the *first talk in a course or session*, you might consider shortening your talk by five minutes. Even in the best organized events, it is not uncommon for the event to start five minutes late. Whenever you start, be aware that some attendees will show up late. Do not make any comments about this and, above all, do not let it affect your concentration.

If yours is the *last talk in the course or session*, you can also shorten your presentation considerably. Despite the chairperson's best efforts, sessions inevitably get behind schedule. Accept the fact that many congress attendees will not show up and some will probably leave in the middle of your talk. Again, you should not make any comments about this and make sure it does not affect your concentration or interrupt the flow of your presentation. The remaining audience is likely to be tired and looking forward to wrapping up, so let them know in your introduction that the presentation has been shortened and you will be finishing shortly – they will appreciate it.

If you are *scheduled to speak just after lunch*, use your knowledge of physiology to your advantage. Have a frugal lunch to prevent a large proportion of your blood from going to the portal system – you will need the full supply of blood to your brain! Remember that the audience will probably have indulged themselves more than you have, and some of them will probably fall asleep during your talk. Do not take this personally, it is merely a question of cerebral hypoperfusion and does not warrant any comment, unless they snore loudly and disrupt your presentation. On the other hand, you should do your best to make your presentation more interesting to keep their attention. In particular, speak a little louder than usual and avoid talking in a monotone.

If you are *scheduled to talk just after the "star" of the program*, in your introduction you should praise the speaker before you and make a humble statement about the content of your own talk. "I'm afraid it will be difficult to arouse your interest in what I have to say after that outstanding presentation by Dr. Foreman" or "I certainly enjoyed Dr. Foreman's outstanding presentation. It is an honor to speak after her, although I'm obviously in another league, I hope I can keep your interest." "That certainly is a tough act to follow, but I will do my best not to bore you". In this situation, it is also probably a good idea to shorten your talk. Again, you must accept the fact that many people will leave the room before your talk or just as you are getting underway. Do not let this disturb you and do not comment on it.

If you are *scheduled to talk just before the "star" of the program*, you might also comment in your introduction that you have been given a bad time slot. Something like "The only advantage of speaking before Dr. Harrison is that I have a reserved seat in the very first row after my presentation." You might also shorten your talk a bit and mention that you are going to be as brief as possible, because you, like everyone else, are looking forward to hearing what the star has to say. Accept the fact that many attendees will enter the room in the middle of your talk and do not make any comments about it or let it undermine your confidence.

If *your talk is the star attraction of the conference*, you are under a lot of pressure to perform. However, if you find yourself in this position, you will probably have earned this right after many successful publications and you will probably be used to giving presentations in English. In any case, be sure to enjoy yourself and see to it that the audience enjoys themselves too. Keep the number of slides to a minimum and do not get bogged down in details. Focus on fundamental concepts; the attendees are looking forward to grasping your overall vision of the topic, not minor details that one of your associates could convey. Take full advantage of this opportunity to sell the best possible image of yourself. Invite the audience to send you their feedback and participate in your research projects, you might be surprised how many will take you up on the offer.

Research

The preliminary work for the presentation of a paper is extensively covered in the literature and is outside the scope of this guide. Here, we just want to mention a few practices that are especially helpful for non-native-English speakers preparing a presentation in English.

Gathering Information

When you gather information for a presentation that will be delivered in English, you should concentrate on English-language sources. Every bit of information (vocabulary, expressions, and ways of commenting on details about your topic) that you have acquired in English is an invaluable resource that can be used in many different ways. If you follow this simple advice, it will be much easier for you to talk about a specific topic in English in public.

Reading from Articles... Aloud

This is a very useful step in the preparation of any presentation in English. You will surely read many articles and book chapters when you are researching your talk. Naturally, you will focus on the contents, but do not pass up the opportunity to practice the pronunciation of the words and sentences you will need to deliver your talk. Many pronunciation mistakes can easily be avoided if you read aloud from the articles you use to prepare the presentation. Reading scientific literature aloud is a good way to identify and subsequently avoid pronunciation mistakes during the delivery of a scientific paper.

Audiovisual Research

Physicians working in academic environments will be familiar with the basics of online and library searches of the written literature. But did you also know that the internet contains a wealth of audiovisual material related to specific medical problems? Two general innovations are of special interest. The first, podcasting, allows you to download audio files to your computer or iPod. You can find relevant podcasts by typing "podcast", "webcast", "vodcast", or "webinar" and a relevant search term, for example "bronchiectasis" into a search engine (e.g. Google, Yahoo!). You will be surprised how many hits you get. The second innovation is online video. Searching is as easy as selecting "video search" in your search engine and entering relevant search terms.

While you are unlikely to find actual sources of information for your presentation in this format, there is a very good chance that you will find material that will be of incalculable value from the linguistic viewpoint because it will give you the chance to listen many of the terms and expressions you will need to say in your talk. Both audio and video files can be downloaded to your MP3 or MP4 player, so you can listen or watch whenever and wherever you like.

Finally, you can watch many scientific presentations on the internet – doing so will help you to appreciate different aspects of presentations outlined in this book that you need to work on. Appendix 6 lists some useful internet resources that will help you research the language you need for presentations as well as improve your presentation skills and general and medical English.

Discussing the Topic of Your Presentation

From the very moment you decide to present a paper in English, you should seize any opportunity to talk about the topic of your presentation with anybody that can contribute ideas or advice – colleagues and English teachers can be especially helpful. These opportunities may not present themselves; in that case you have to create them.

Looking up Both the Meaning and Pronunciation of Unknown Words in the Dictionary

We often look up the meaning of the words in the dictionary but fail to take full advantage of the time devoted to this task because we do not double check the pronunciation of the word. Today it is not even necessary to understand the phonetic alphabet or other signs used to convey sounds in dictionaries – many online or CD-ROM dictionaries allow you to listen to the pronunciation with a click of the mouse. Appendix 6 also lists a few online sources that will allow you to hear the words you need to know pronounced by a native speaker.

Latin and Greek terminology can be tricky. Although these words are generally easy to understand because they form part of the medical vocabulary in many languages, each language has its own system of pronouncing these words and English is no exception. Appendix 1 gives a brief discussion of the rules for pronouncing these words in English, but there is really no substitute for hearing a word spoken by a knowledgeable native speaker and the above-mentioned dictionaries are ideal for this purpose.

Jotting down Tricky Words and Sentences

Write down key words and tricky terms as you come across them. This can help you with both pronunciation and spelling.

Organizing Your Ideas

Only a tiny fraction of the information gleaned from your research can be included in your presentation. It is natural to want to include all of the relevant interesting points that you have picked up along your way, and that partly explains why inexperienced lecturers tend to try to cram as much information as possible into their allotted time. You will soon realize that this approach is unviable and you will have to make some hard choices about what to include and what to omit. More than anything else, what you decide to include and to leave out will determine the character of your presentation.

While time constraints limit how much you can say, you control what you say by adjusting the two dimensions that delineate any treatise: scope and depth. Narrowing the scope of your presentation will enable you to go into greater depth; on the other hand, broadening the scope of your presentation can allow you to avoid depths that you audience would not be able to fathom. It is essential to consider the makeup of your audience when you make these decisions. You must strive to reach the right level of depth in your presentation, and you will succeed or fail at this task depending on how well you have judged your audience's knowledge about your topic. You cannot expect an audience of medical students and residents to understand a talk that is directed at other experts in your subspecialty.

Surprisingly, the vast majority of lecturers make the mistake of assuming that their audience knows more than they possibly could about the topic of the lecture. Whether from a desire to impress other experts, a fear of being simplistic, or a merely failure to remember what it is like to be starting out, they fail to recognize the enormous differences between their knowledge and their audience's and consequently they fail to communicate their ideas.

Making the opposite mistake – presenting ideas that are too simple – is rarely as disastrous, as long as it is well done. Even experts can enjoy a well-prepared review of relatively basic material if everything else is in order. But beware – if you give too simple of a talk to a group of people well versed in a topic you had better get everything right or be prepared to be challenged.

Most audiences are mixed, so it is best to include a little something for everyone. Do not neglect to include essential background material early in the presentation to make sure you do not lose your audience at the starting gate; if you were to do that, they would never have a chance to follow you through the rest of the presentation. Do not hesitate to dive briefly to depths where only other experts could follow you, but the key word here is briefly. Any extended forays into these areas will likely alienate the rest of the audience and it might be difficult if not impossible to reconnect with them at a more superficial level. Whenever, you go deeper than a large part of the audience can follow, it also helps to summarize the more involved, technical material by restating it in simplified terms, using expressions like "in other words" or "by analogy"; this keeps those who cannot follow you from drifting off.

It is important to orient the audience from the very beginning of your talk. First, it is essential to provide them with the information necessary to enable them to understand the exact subject of your talk, to decide whether it is important to them, and to have some idea of the background that will be necessary to follow your presentation. Then, you need to give them an outline of the major points you will be dealing with and the order you will tackle each of them. Do not be afraid of "wasting" too much time in mapping your talk, this initial information will go far toward making the rest of your talk comprehensible. The old adage "tell them what you're going to tell them, tell them, and tell them what you've told them" works well for most presentations.

Organize the body of your talk to ensure a logical narrative with smooth transitions between points. Always make sure that your audience knows where you are leading them. This is even more important in a presentation than in a written document, because, unlike readers, the audience cannot go back or stop and ponder where you are taking them. Signal changes of direction or moving on to new subject new subject matter clearly, using both visual cues (specific slides) and delivery (pauses, inflection).

Finally, audiences want closure. You should try to summarize the main point of your talk into a single sentence. Your last slide should list no more than three or four main conclusions, and you should allow ample time for you to go over each one. This is where you can hammer home the points you made in the rest of the talk – don't worry about being repetitive – repetition is one of the most important tools you have to work with. Unless you emphasize your main points by repetition, nobody is likely to remember them.

Chapter 3 provides specific advice about language points to help you with different parts of your talk and transitions between them.

Creating Your Slides

To a large extent, your slides will determine the success or failure of your presentation. Your slides will provide an outline for both you and your audience to follow throughout the talk; thus, the organization and clarity of your slides is crucial to ensuring that your message gets through to the audience. Your talk should have a beginning, a middle, and an end; it is essential for the audience to know where they are in this structure at all times. One way to accomplish this is to provide listeners with a map of your talk near the beginning. It is always a good idea to let them know what is to come. To communicate effectively, your slides should be arranged to explain a narrative.

However, you must never forget that slides are not the main vehicle of communication in a presentation. Slides should support and reinforce what you communicate to the audience through your speech; they should never distract the audience's attention from the main message that you are trying to communicate with your voice.

Each slide should fulfill three inclusion criteria: it should be error-free, simple, and necessary to your narrative. You should make every effort to ensure that the mechanics (spelling, grammar, and punctuation) on your slides is perfect. Whereas listeners might be willing to forgive small mistakes in the spoken part of your presentation, there can be no excuses for mistakes in your slides. Be sure to use your computer's spell check program, and ask others to pay special attention to any possible errors in your slides while you rehearse. Chapter 4 discusses common mistakes in English language mechanics.

Limit the information on each slide to a single important point and its supporting material. Cluttered slides are difficult to read and only serve to confuse readers. Simplify graphic material if necessary to make it immediately comprehensible. You can always fill in the details with what you say.

Ask yourself if the slide is really necessary. Is it truly relevant to the main narrative or is it sidetracking? How does it fit into your narrative? If the answers to these questions are not immediate obvious to you, you cannot expect your listeners to understand why you have included this material.

Finally, there are some formal aspects of slide preparation that deserve mention. Experts have studied the effectiveness of different aspects of slide design, and we would be foolish to ignore their findings. Here we provide just a few guidelines for designing effective slides.

Number of Slides

As a rule of thumb, you can plan on about one slide per minute if you are an experienced lecturer and about one and half slides per minute if you are a beginner with fewer resources to fill the gaps. When you rehearse your presentation, you will see if you can add a few slides or leave a few out.

Trying to present too many slides is one of the many pitfalls that can spoil your presentation. You don't want to have to fight against time to get through all of your slides.

Do not forget that when you deliver a presentation, you are not supposed to be in a rush.

Number of Lines per Slide

Another pitfall is trying to cram too much information onto an individual slide. You should limit the number of lines to six, including the headings. If that seems like very little, remember that the slide should only give the key points you want to reinforce, the bulk of the information you want to include will be found in what you say (that's why it's called an oral communication). You should never make the terrible mistake of presenting a slide in which you merely read what is projected to the audience or you will run the risk of insulting and angering your listeners.

Graphic Material on Slides

Diagrams should be simple and uncluttered. Figure legends for medical images should be absent or minimal. Again, you should fill in the gaps in the visual information with your spoken words. Choose the best kind of graph to display your data, and make sure that everything is clearly labeled. Do not post in one slide what can be posted in two. Chapter 3 provides some examples of how to use graphical material in slides.

Fonts

A boldface sans serif typestyle like Arial is easiest to read. Avoid using all capital letters, as this slows down perception considerably. The size of the letters will depend on the number of rows of the conference room. In case of doubt, the bigger the better.

Background of the Slides

Contrast is essential: you should opt for either a plain dark background with white or yellow text or a plain light background with very dark text. Color and design should never distract from the principal texts and figures. Avoid green on blue as color-blind individuals will be unable to see the contrast.

Rehearsal

Once you have decided what you are able to say in the amount of you time you have been allotted and have prepared the slides that will serve as the skeleton of your presentation, it is essential to rehearse your delivery. In fact, you will discover that you cannot actually be sure of what you want to say until you have rehearsed your talk a few times. Rehearsing allows you to check the timing of your presentation, ensure smooth transitions between slides and between sections, and to discover the weak points in your presentation while you still have time to correct them. You will see how rehearsing gives you confidence, because being prepared is the best safeguard against nervousness.

We will briefly discuss some aspects of delivery that we can improve by rehearsing and give you a few tips to help ensure that you make the most of the limited time you have available for rehearsing.

Writing Out Your Speech

Some presenters find that it is useful to write out the entire body of text that they want to say during their talk. This is not a bad idea, but you must be careful because what sounds good in writing will not necessarily sound good in speech. Be sure to use short sentences with simple language. After you write out your talk, read it out loud to see whether it sounds natural and whether you feel comfortable with the language. You will probably have to change many things to arrive at a text that you feel comfortable with.

Reading Your Presentation

Reading the text of your presentation from paper or, even worse, off your slides should not be considered a valid option. In fact, you should not read any slides in your presentation. Reading your slides aloud to the audience undermines your credibility as a lecturer and insults your audience's intelligence; after all, it is safe to assume that they all know how to read. Remember that your slides should help orient your audience to the message you deliver orally; slides cannot take on the burden of delivering your entire message.

However, if your English is so bad and you are so nervous that you cannot possibly imagine doing anything but read your presentation, here are a few tips. Try to make it as natural as possible – look up from your paper as often as possible and try to make eye contact with some of the audience. Make sure that your text is as simple as possible – use short sentences with simple language. It may help to imagine you are writing your text for an intelligent layperson rather than for an expert in your field. Rehearse as much as possible, be careful not to read too fast, be sure to pause where appropriate, and avoid speaking in a monotone.

Memorizing Your Presentation

It is not usually a good idea to memorize the exact wording of your presentation, because this practice entails the risk of getting lost. Some of the most surrealistic situations witnessed at international scientific presentations have been related to "shy lecturers" delivering "memorized talks" going blank under pressure. Moreover, memorizing all but the shortest of scientific presentations would require enormous investments of time that most busy professionals cannot allow themselves.

However, this does not mean that there is no role for memorization. It can be useful to memorize key parts of your talk like the introduction or take-home points as well as difficult parts that have given you trouble during rehearsal. Moreover, it is essential to memorize the IDEAS of your presentation. Your slides will help you and your audience with an external outline, but you also need to have an internal outline at your disposal when you get up to give your talk.

If you rehearse your talk often enough, you cannot help but memorize a large part of the text. If your English is not very good, you will need to rehearse some of the phrases many times to be able to pronounce them clearly using acceptable intonation. The greater your proficiency in the English language, the more resources you will have to allow you to concentrate on the ideas of your talk and the less you will need to memorize exact word combinations. Consequently, proficient speakers have a much lower risk of stumbling over a point if they forget the wording they intended to explain it.

Timing

One critical point that can only be dealt with by rehearsal is timing. You can never be sure how long your presentation will take until you have clocked yourself. You may discover that you have to alter your presentation to fit your time allotment; that is why it is important to finish putting your presentation together well before your speaking engagement. Remember that most people have a tendency to speak faster when they are nervous, and plan your talk to be a bit shorter than scheduled – both organizations and audiences have little patience with lecturers that run over time, and too many presenters are cut short by the chairman's words: "Dr. X your time is over". One trick you can use to avoid running over time is to prepare a three-minutes conclusion slide that you can jump to from any point near the end of your presentation. If you haven't finished presenting your details when you are three-minutes from your time limit, you can jump to this slide and say something like "If we continue along these lines, we will reach these conclusions". Clearly, this strategy can only be considered as a safeguard against a disastrous ending and cannot replace careful planning and rehearsal to get the timing right.

Pacing

Timing depends on pacing. While in general it is best to speak slowly and clearly to make sure that your audience can follow you, too slow of a pace will lull your audience to sleep. In a good presentation, your pacing changes constantly to engage the audience. Slow down when you are explaining complex or difficult material. Pauses are important: you should pause after introducing an important point and after introducing a new image or graph.

Pronunciation

Even native speakers can have difficulties pronouncing some words. The only way to find out which words you tend to stumble over is by rehearsing. Although some words can be replaced by others that you find easier to pronounce, medicine is full of long words that can tie anybody's tongue into knots, and although it is often possible to find a replacement term, it will probably be necessary to say these words at least a couple of times. One example of a word like this is lymphangiomyomatosis. If you were doing a presentation on this topic, you could use the abbreviation LAM in most cases, but you will have to pronounce the full word at least twice (once at the start of your talk and again at the end). Our advice for learning the pronunciation of words like this is:

1. Divide it up in its etymological components (lymph-, angio-, myo-...)
2. Check that you know the meaning of all these components
3. Read these components aloud separately
4. Read them in pairs "lymphangio", "myomatosis"
5. Read the entire word aloud
6. Say it as many times as possible, for example, on your way to the hospital.
7. Try to talk to your colleagues about this word asking them for instance: Have you noticed how difficult it is to pronounce "lymphangiomyomatosis" in English?
8. In short, make the tricky term an easy one by repeating it over and over.

Weak English speakers will need to work hard at their pronunciation. Rehearsing with a native speaker or very fluent English speaker can help to make sure your audience will understand you. Ideally, you should work with a native speaker who has some idea of linguistics, because word stress and sentence intonation are essential to comprehension. Even fluent English speakers will mispronounce some words – words that are written the same in English as in your mother tongue can be especially problematic. The first step is to find out which words you mispronounce – most speakers find that once they become aware that they tend to pronounce a particular word wrong, this problem corrects itself.

Chapter 10 gives more advice on how to improve your pronunciation.

Eliminating Filler Phrases

Many people tend to fill gaps in their speech with empty phrases like "you know" or "basically" or with sounds like "uhm" or "er". To a certain extent, this is only natural and throwing in the odd filler phrase from time to time to allow yourself time to think will probably be tolerated by most members of the audience. However, excessive use of filler phrases can be extremely irritating to some members of the audience. Listening to a recording of yourself will show you which filler phrases you tend to use and also point out where you tend to need them. Like words you often mispronounce, filler phrases tend to correct themselves once you become aware that they are there.

Presenting the Paper to Yourself

Once you have created your presentation, you should first deliver it to the most demanding audience in the world: yourself. Talking in front of a mirror is a good first approach. As you watch yourself, be gentle with your criticism and never cease imagining yourself giving a successful presentation. You can practice your presentation in front of the mirror every time you shave or put on your makeup. Although you might not have time to rehearse the whole presentation in these situations, you can rehearse key parts like the introduction, conclusion, or transition slides.

When you get to the point where you start to know the talk well, it is a good idea to make a video recording of the entire presentation. You can learn a lot about yourself (your movements, expressions, voice, and pronunciation) and about your presentation (strong points, weak points, and incoherent points) by watching this recording.

Rehearsing the Weak Points Alone

By rehearsing, you will soon learn where the weak points of your presentation are. Concentrate on rehearsing these weak points. Although you should repeat the entire presentation a few times over the course of your rehearsal to make sure that you get the timing right and to become comfortable with the whole package, there is no need to waste time repeating the whole presentation every time you rehearse. In fact, doing so, you run the risk of becoming sick and tired of your own presentation. Remember that if you don't like your presentation, the audience is not going to like it either.

Presenting the Paper to Your Colleagues at Your Department

Delivering your presentation to your colleagues in a clinical session is a great way to get constructive criticism and positive feedback. Invite other English-speaking professionals from outside your department, too, as they will see things from another point of view that will enable them to give you priceless advice. If nobody in your department speaks English, you can deliver your presentation in your own language – this will help you become familiar with your slides and your topic. Although you will not be able to rehearse the pronunciation of the talk, at least you will check its spelling and get started with your presentation.

It is important to do this "dress rehearsal" early enough to allow you to fix anything that your colleagues pointed out to you that is not as good as it could be and to incorporate suggestions. It would be pointless and a waste of everybody's time to wait to do this important step the night before your talk.

Final Preparations

Chapter 5 gives some advice on how to deal with some of the most common problems that can occur during your presentation. It is always wise to imagine and rehearse dealing with some of the many things that can go wrong during your presentation.

Few lecturers would be able to deliver their talk without any visual support and you must take every step to ensure that your presentation slides will work when you need them to. For that reason, you should record your presentation on at least two different media, for example on a CD and a USB data pen. It is also a good idea to store the presentations in different formats and bring along any unusual fonts or plug-ins that might give you problems. As a final precaution, it cannot hurt to e-mail the presentation to yourself at a web-based e-mail address (gmail, yahoo!mail, or hotmail) as an attached file – if all else fails, you will be able to recover your slides from any computer with an internet connection.

Last minute rehearsal is not always possible and is seldom productive. You would take better advantage of your time enjoying the company of your colleagues and building social networks than locking yourself in your room to run through your talk once again. Try not to eat or drink too much and make sure you get a good night's sleep.

On the day of the presentation itself, you should also be careful not to eat or drink too much or too little – you don't want to have to deal with additional physical discomfort like a growling stomach or the urge to urinate while you're behind the podium. Some lecturer's take a minimal dose of beta-blocker to break the negative feedback loop by eliminating external signs of nervousness like trembling or sweaty palms, though we suspect that the efficacy of this measure is mostly due to the placebo effect. In any case, if you resort to beta-blockers, it is probably a good idea to try them out on another occasion before your big day. Chapter 5 gives additional advice on dealing with nervousness.

Be sure to dress appropriately. For most scientific talks, this means formal, conservative clothing, i.e. a suit (preferably dark) and discreet tie for men and a dress or pant-suit for women. It is also important that your clothing make it easy to wear a portable microphone. The microphone should be attached to the side of your jacket, shirt, top, or dress that is nearest the screen so that the audience can hear you when you turn toward the screen to point something out. If you have any doubts, ask the organizers and follow their advice.

CHAPTER 3

Chapter 3 Useful Language for Scientific Presentations

The specific language you need for your presentation will obviously depend on your particular topic. You will learn this specific vocabulary as you prepare your talk. In this chapter, we present some general expressions that can help you deal with certain aspects and situations common to most presentations together with a few concrete examples of the different approaches applied to specific situations.

Opening Your Talk

Good morning/afternoon/evening, ladies and gentlemen. Remember that in English the morning lasts from the beginning of the session to 12 p.m. (noon). The afternoon comprises the period from 12 p.m. to sunset; however, many speakers say "good evening" from about 6 p.m., regardless of the time of the year.

It is customary to thank the moderator, chairperson, or person that has introduced you.

Thank you, Dr. Ross, for your kind introduction. You can get by with *Thank you, Dr. Ross* or even just saying *Thank you* and looking at the person who introduced you.

If nobody has introduced you, briefly introduce yourself.

My name is Susanne Godin and I'm a neurosurgeon at La Pitie Hospital in Paris, France.

If you are an invited speaker, you should thank your host.

I would like to thank Dr. Hansen and McGill University for giving me this opportunity to talk to you today.

You will probably have sufficient time to include a few words of praise for your hosts, their institution, and/or city or country.

It is an honor to speak in such a renowned institution.

Montreal is a great city and it's great to be here.

Mention any connection you might have with the institution or city.

I did a fellowship here in Dr. Wong's lab back in the 90s.

This is my first visit to Berlin; I hope it is not my last.

Make sure your audience understands the subject of your talk.

Today/this morning/this evening/in this presentation, I'm going to tell you about….

J. Giba and R. Ribes, *Preparing and Delivering Scientific Presentations,*
DOI: 10.1007/978-3-642-15889-6_3, © Springer-Verlag Berlin Heidelberg 2011

It is a pleasure/I'm very grateful/to have this opportunity to talk to you about….

It is sometimes useful to inform them what will NOT be included in your presentation.

This presentation will not give you a general overview of the complications of hip replacement surgery; rather we will limit our discussion to prosthetic loosening.

I'm afraid that we won't have time to delve into the interesting new developments in minimally invasive surgery, but I understand Dr. Anderson will be talking about that later today.

It is also good form to acknowledge your colleagues during the introduction.

I'm presenting this paper on behalf of my colleagues, Drs. Beluga and Young.

Dr. Casey and I first became interested in this subject after…

I am grateful to have the support of experienced, knowledgeable researchers like Sarah O'Reilly, Larry Peterson, and Bill Edwards, as well as of brilliant and hard-working research fellows like Jane Goodman and Dana Smith.

In the following sections, we present some sample slides with brief examples of what the presenter might say while each is displayed. Of course, what the presenter says will obviously be determined by the purpose of the talk and by the target audience. Experienced presenters often include the same slides in different presentations with completely different explanations, depending on their intended audience.

Mapping Your Talk

It is often a good idea to include a "mapping slide" early in your presentation and repeat it, highlighting the relevant part, when you make major transitions. Here is one example of a mapping slide from a presentation about the clinical value of antinuclear antibody determinations:

Contents

- What are antinuclear antibodies (ANA)?
- Tests to detect ANAs
- Indications for ANA testing
- Interpretation of test results
- Conclusions

This slide was displayed immediately after the title slide. It provided an overview of the entire talk and also served as the introduction to the first section. Later in the talk, the same slide was displayed at the beginning of each section with the relevant title in black and the remaining sections in gray. The following slide was displayed during the transition to the second section of the talk:

Contents

- What are antinuclear antibodies (ANA)?
- Tests to detect ANAs
- Indications for ANA testing
- Interpretation of test results
- Conclusions

This slide was followed by another that provided an outline to the second part of the talk:

Tests to detect ANAs

- Indirect immunofluorescence
- ELISA
- Substrate
- Fixation
- Microscopy

In this approach, the section title is often used as the heading for all slides within a section.

Introducing the Main Question

After providing your audience with all the necessary background information they need to understand why the matter you discuss is important, you need to zoom in on the precise subject matter of your presentation. The following phrases can be useful for introducing your objective, hypothesis, or research question:

- *The question is then ….*
- *This made us wonder….*
- *So, the question is this ….*
- *Thus, we aimed to ….*
- *Therefore, we hypothesized that….*
- *So, we wanted to know….*

Talking about Methods

After introducing your research question, you will need to tell your audience how you went about answering it. Although some presentations describing new and innovative research techniques may dwell more on the methods than any other section, the aim of most short presentations is usually to communicate your findings and results, so you should limit discussion of the research techniques to only the essential details.

- *These were the inclusion criteria:*
- *We excluded patients with any of the following:*
- *We recorded the following variables:*

You may have to mention any special equipment you used. A slide with a picture (a photograph or diagram) of any device your audience may be unfamiliar with and interested in can help (see below).

- *This is the quadrature head coil we used for the MRI studies.*

Standard techniques can be mentioned briefly in passing.

- *We used ELISA for all determinations.*
- *All patients underwent T1- and T2-weighted MRI sequences.*

It may be necessary to define some variables before moving on to the results.

- *We considered prostates with asymmetries, indurations, or nodules abnormal.*
- *TIA was defined as an isolated episode of amaurosis fugax or focal cerebral dysfunction of ischemic origin with complete recovery within 24 h.*

A flowchart can often provide useful support for talking about the patients or subjects included in the study. As with other figures, it is important to keep complexity to a minimum and to clearly point out details that you want the audience to pay attention to. Building up the chart point by point can help the audience focus on the relevant parts of complex flowcharts.

Presenting Data in Tables

It is often convenient to present data in a table. Remember, however, that a presentation is not a written report. Whereas readers can study tables at their leisure, the audience at a presentation does not have this luxury. Ask yourself whether the information in the table is strictly necessary and simplify tables that you cannot eliminate. Consider presenting the data in another format (e.g., a graph) that will be more amenable to your audience. Pause when you display a new table and be sure to give your audience time to take in the information it contains.

Try to design your tables so that they can be understood without any accompanying explanation. Every table should have a brief title to inform the audience of the specific topic or key point that it reports. Label the columns and rows clearly. Define any nonstandard abbreviations you need to use. Give units of measurement (usually within parentheses) beside or below the variable in the head of the column. If necessary, add footnotes with some of the following information: details of the experiment, definitions of abbreviations and/or symbols, information about statistics.

Even after your best efforts to design a table that is perfectly comprehensible on its own, you still need to lead the audience through the table, directing their attention at the key points you want to stress. Remember that while you may know everything in the table by heart, this is probably your audience's first encounter with this information and it is your job to explain it to them. A lot of data can be crammed into a table, and it is the presenter's job to point out what is relevant. Mark the key points clearly, so your audience does not have to struggle to determine what you are trying to say. It is a good idea to design your slides so that the information you want to draw the audience's attention to is clearly marked, so you do not need to actually point it out with the computer's mouse or a laser pointer, and it is often preferable to use effects to highlight the important points one by one as you talk about them.

FETAL GROWTH

Week	**8**	**12**	**16**	**20**
Weight (g)	2	17	142	340
Length (cm)	2.5	7.5	16	25

This slide shows how the fetus grows from the second month of development to term. I'd like to remind you that we're referring to developmental age here, in other words, to the time from fertilization. Gynecologists usually refer to gestational age, which is calculated from the mother's last period and is typically 2 weeks ahead of developmental age. Thus, a developmental age of 8 weeks corresponds to a gestational age of 10 weeks.

During the third month of development, from the 8th to the 12th week, the fetus grows phenomenally, tripling its length and increasing its weight eightfold (here you should point out the relevant figures in the table).

Sometimes it is a good idea to incorporate visual or graphic material into a table to make it easier for the audience to grasp the main idea. Notice how a graph would illustrate the same text somewhat better:

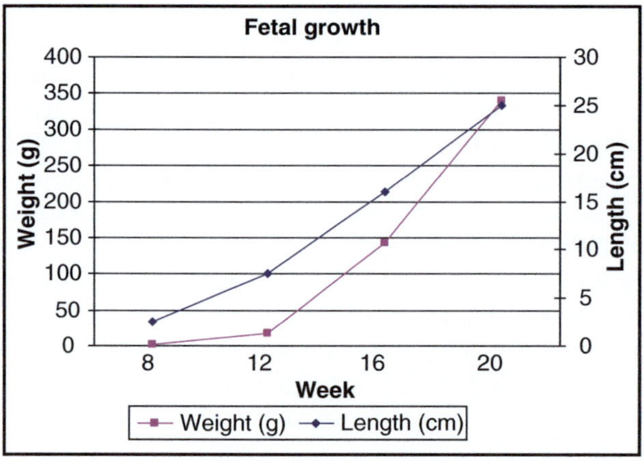

But perhaps the best way to make the data come alive is by adding an image to the table:

FETAL GROWTH

Week	8	12 	16 	20 
Weight (g)	2	17	142	340
Length (cm)	2.5	7.5	16	25

Talking about Figures

A picture is worth a thousand words. Like tables, figures should have a brief title to inform the audience of the specific topic or key point that they illustrate. You should pause briefly every time you display a new figure so that your audience can take in the information it contains. Presenters are often so familiar with their material that they forget how difficult it is for someone seeing the information for the first time to get oriented.

Try to design your figures so that they can be understood without any accompanying explanation. Label the image clearly. Define any nonstandard abbreviations you need to use. Give units of measurement (usually within parentheses) on the axes of the graph. If necessary, add footnotes with some of the following information: details of the experiment, definitions of abbreviations and/or symbols, information about statistics. However, remember that the bulk of the explanation should be contained in what you say – figures already contain a lot of information for the audience to process, and you do not want to overwhelm them with an image cluttered with too many labels.

No matter what kind of figure you display, be sure to use the same key words in the figure as in the rest of your talk. For example, if you have referred to the sulcus lateralis cerebri as "the lateral sulcus" throughout the talk, you could confuse some listeners if you label it "the Sylvian fissure" in an image.

General Language for Referring to Images

Although it is usually best to use a pointer (or even better – previously placed arrows or other markers) to refer to important items in a picture, it is sometimes useful to direct the audience's attention to different parts of the picture in general terms. It is possible to describe just about any position within an image by combining the information presented in the two following slides. It is especially important to use the correct prepositions when you use this approach.

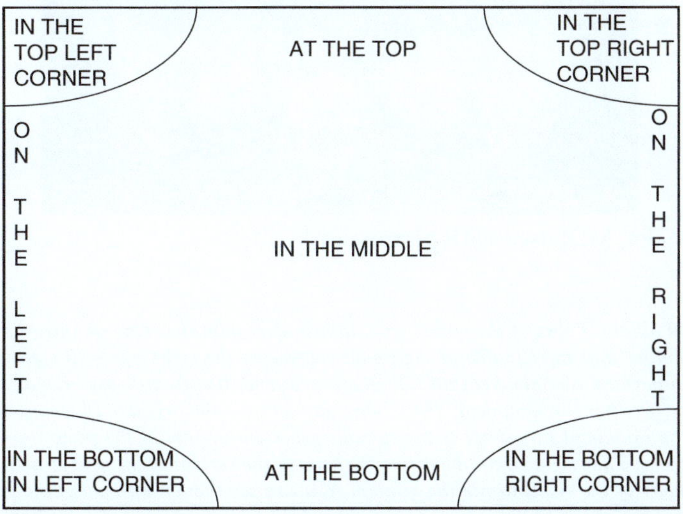

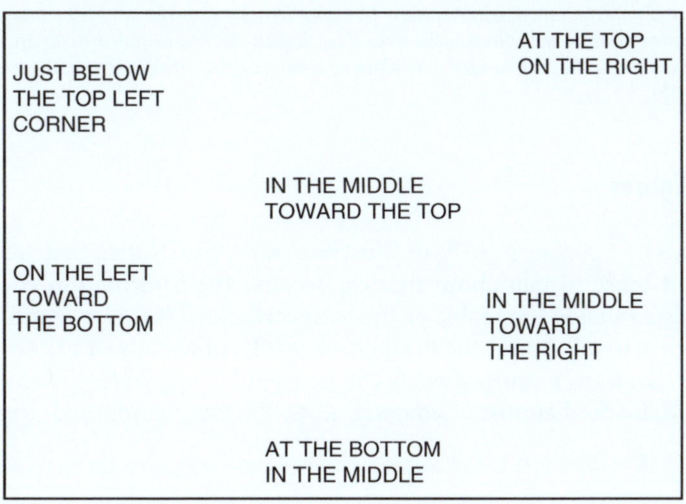

This kind of language is often used together with prepositions that describe the relation between the objects in the picture, like *on*, *in*, *under*, *above*, *to the right/left of*, *near*, *behind*, etc.

Here is an example of how we might use this kind of language in a talk to teach basic radiological anatomy to medical students:

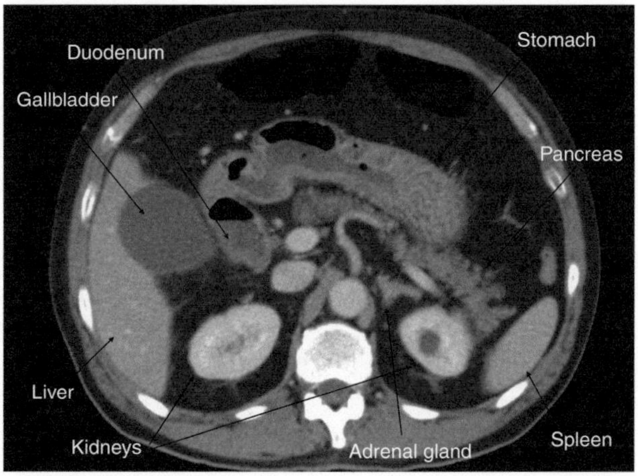

(Antoni Malet, MD; used with permission)

This is an axial CT slice of the abdomen obtained after administering an intravenous contrast agent. Remember that the left side of the image represents the right side of the patient and vice versa. It is important to remember that CT images are generated from X-ray images, so the denser the tissue the brighter it will appear. Thus, bones are white – the bright Y-shaped structure at the bottom in the middle of the image is the spinous process and part of the posterior laminae of a lumbar vertebra. The bright circle above it is the body of the vertebra. The oval structures on either side of the body of the vertebra are the kidneys. You can see the left adrenal gland above the left kidney toward the center of the image. The slightly darker, more elongated oval shape to the right of the left kidney in the image is the spleen. The long structure on the left of the image that has a density similar to that of the spleen is the liver. The slightly darker, rounded structure abutting the top half of the liver is the gallbladder. The black spaces at the top of the picture illustrate air in the intestinal loops.

Types of Figures

Many types of figures can help us illustrate our ideas. Notice that we often use the present tense to talk about figures, because the information contained in figures exists outside the realm of time. Nevertheless, it is also common to use the past tense to talk about the findings or results of a study or experiment.

Figures that show findings directly can be useful for providing examples: these may come from diagnostic imaging tests (X-rays, computed tomography, magnetic resonance imaging, ultrasonography, scintigraphy, positron emission tomography, echocardiography, endoscopy, etc.), photographs (of patients, of the surgical field, of anatomic specimens, of apparatuses, of results of techniques like electrophoresis, etc.), or printouts from monitoring devices (electrocardiogram, electroencephalogram, etc.). No matter where the image comes from, it is essential to eliminate any labels or facial characteristics that might allow the patient to be identified.

Here is an example of a slide from a presentation entitled "CT Findings in Chronic Pulmonary Embolism":

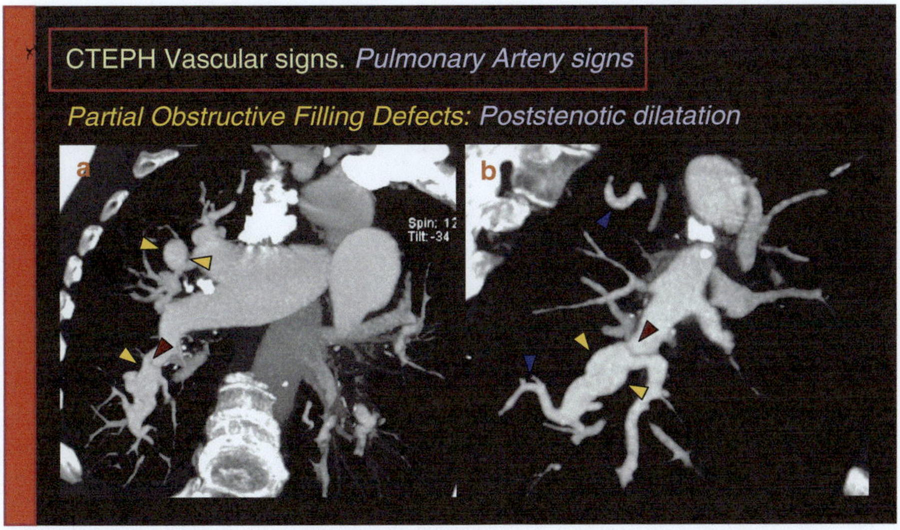

(Eva Castañer, MD; used with permission)

While this slide was being projected to an audience of radiologists, the speaker made the following comments:

These patients often have aneurysms or poststenotic dilatations. In the image on the left, there are two aneurysms – one affecting the posterior segment of the right upper lobe (at this point in her talk, the yellow arrows marking the dilated segment appear) and the other affecting the right lower lobe artery (the other yellow arrow appears). The image on the right is an oblique MIP: here we can see the band in more detail (red arrow appears in both images) and we get a better view of the poststenotic dilatation in the right lower lobe artery (yellow arrow appears in the image on the right). There are also evident signs of PH like marked increase of diameter of pulmonary arteries and tortuous vessels (blue arrows appear).

Here's another example of a figure that shows findings from a patient directly, in this case, an electrocardiogram. This slide was shown in a presentation to a group of general practitioners in the third class in a mini-course on electrocardiography.

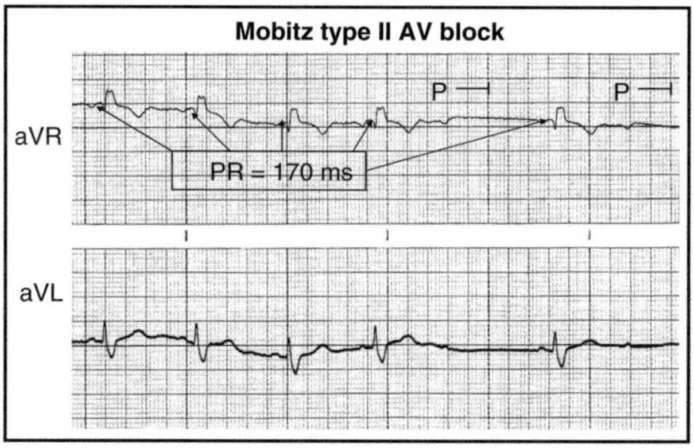

(Antonio Martinez-Rubio, MD, PhD, FESC, FACC; used with permission)

Here's another example of second-degree atrioventricular block. Let's have a look at these two simultaneous EKG traces from the aVR and aVL leads. (Pause) You can see how the first four P-waves are followed by four QRS complexes. The P–R interval is a constant 170 ms throughout these four beats. (Pause) But look what happens next! (Pause) The fifth P-wave is not followed by a QRS complex. (Pause) This happens again after the seventh P wave. (Long pause). What does this mean? (Pause for a few seconds to give the audience a chance to begin to formulate their own answers to this question). Remember the P-wave corresponds to the depolarization of the atrium, and the QRS complex corresponds to the depolarization of the ventricle. (Pause) So, what we are seeing here is an abrupt block in the electrical impulse that travels from the atrium to the ventricle.

This is what we usually refer to as a Mobitz type 2 AV block. Instead of the progressive lengthening of the PR interval, called the Wenckebach phenomenon, that we saw earlier in Mobitz type I, the interruption in P-wave conduction is sudden and unexpected in Mobitz 2, and as we're going to see, this has important implications for the clinical presentation and treatment.

Other figures, like drawings or diagrams, are useful for summarizing and illustrating ideas. Drawings illustrate concrete objects. Diagrams illustrate processes or concepts. Both drawings and diagrams can be more or less realistic or more or less schematic. Two principles should guide your use of drawings and diagrams: simplicity and clarity. Do not present more information than your audience can absorb, and be sure to direct their attention to the relevant parts of the figure both by marking the image and by what you say. A drawing is often better than a photograph because it allows us to emphasize important details while eliminating or minimizing less important details that might distract the audience in a photograph. Drawings also allow us to use cross-sectional cutaways to show details below the surface that are not visible in photograph. Compare the usefulness of this photograph and this diagram as visual support for the spoken text below:

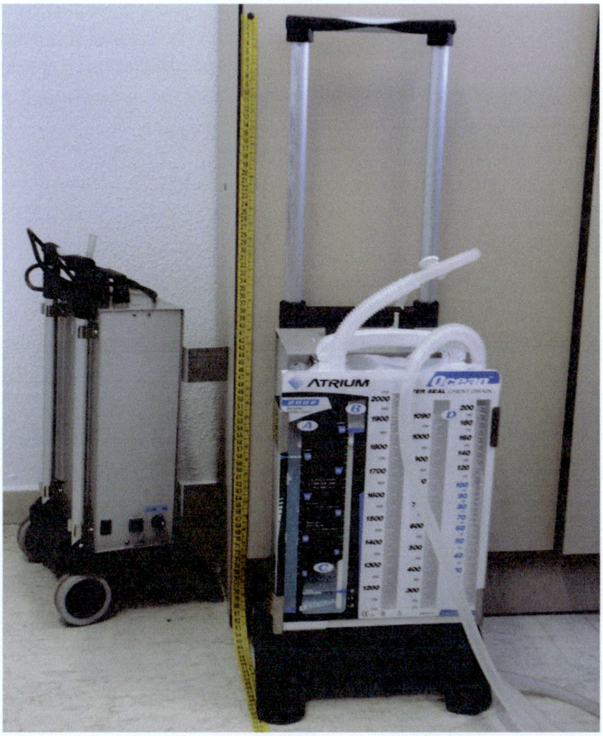

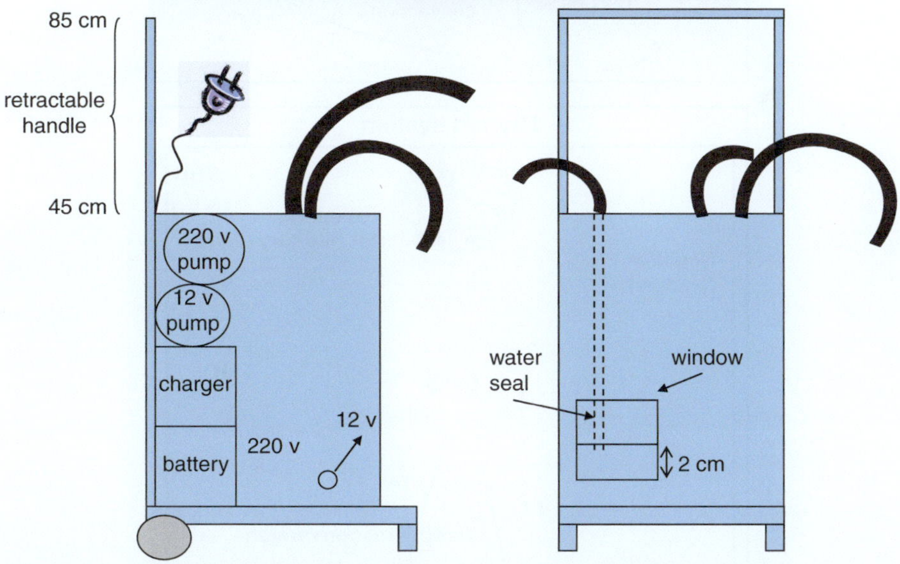

The ADS is an electronically controlled aspiration system to collect fluid/air. It has a water seal that allows the aspiration pressure to be kept constantly above 20 cm H_2O. As you can see, it is equipped with a handle and wheels, so that the patient can pull it around. It has a dual power source. When patients

are moving around, it runs off a 12-V battery. And when they are stationary at home or in their hospital rooms, they can plug into a standard 220-V AC outlet. The ADS has a separate aspiration pump for each power source: so when they are plugged in, the standard hospital aspiration pump kicks in, and the battery charger comes on. Having separate pumps for each energy source is much more energy efficient, extends the battery autonomy, and also allows for faster charging.

Whether you use a more realistic or a more schematic diagram will depend on many factors, such as your artistic ability, your expected audience, or personal taste. Schematic diagrams have the advantage of being simpler, but the visual components of more realistic diagrams often help to make them more memorable. Compare the usefulness of these two diagrams in illustrating the accompanying text about the hypothalamic-pituitary-thyroid axis below.

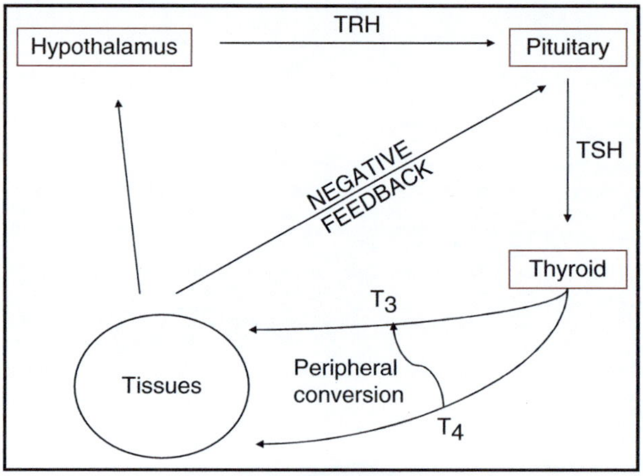

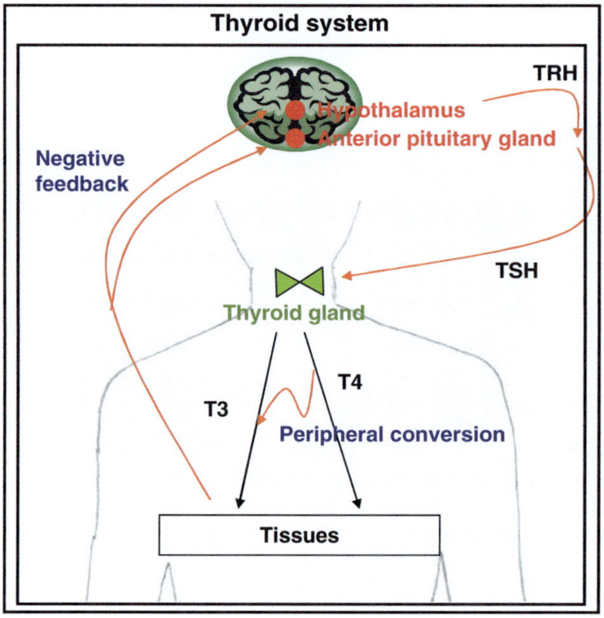

When the hypothalamus senses low levels of thyroid hormone: triiodothronine or T3 and thyroxine or T4 in the circulation, it responds by releasing thyrotropin-releasing hormone, or TRH. The TRH acts directly on the pituitary, stimulating it to produce thyroid-stimulating hormone, or TSH. The TSH, in turn, stimulates the thyroid to produce T3 and T4. T3 is the active form of thyroid hormone. It is much more potent than T4, and about eighty to ninety percent of the thyroid hormone is released from the gland in the form of T4, which is converted to T3 in the tissues. When the levels of thyroid hormone return to normal, negative feedback to the hypothalamus and the anterior pituitary inhibits the release of TRH from hypothalamus and TSH from anterior pituitary gland.

Graphs are figures that are useful for summarizing data and conclusions. Audiences can often appreciate information presented in graphs more intuitively than they can information presented in tables. There are many different kinds of graphs, and it is important to choose the one that best suits your purposes. Bar graphs are useful for comparing differences in whole quantities. Pie graphs are useful for showing the different proportions of each element in relation to the whole. Boxplots are a convenient way to show a great deal of information about groups of numerical data, but they can be confusing for many audiences and you should take your time explaining them. Line graphs are most useful for displaying data that changes continuously over time. By convention, the independent variable is plotted on the X-axis and the dependent is plotted on the Y-axis.

Here is a very simple example of a graph, in this case a histogram, from one of my classes in which I talk about the importance of English for communication in science:

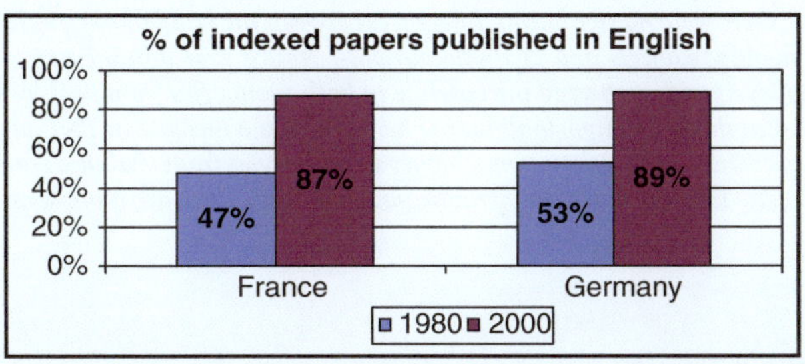

As I put up the graph, I tell my students what it is about; for example, "*this graph shows the difference in the percentage of papers published in English in France and Germany in a twenty-year period*". But I don't stop there – I know I need to go on to explain the different axes and bars. "*The blue bars show the percentage of papers that were published in these countries in 1980, which as you can see, is roughly half of all scientific papers. The red bars show the percentages for the year 2000, which is nearly 90%.*" Then I tell them the important point "*As you can see, English is becoming more and more important for scientific communication*". Finally, I try to make it clear how this information relates to the rest of my presentation "*Together with the other figures we have seen, this example shows us that the trend toward communicating in English is becoming more and more pronounced, and although I don't have the figures, you can bet that this percentage will have increased by 2010.*"

The following slide is an example of how results might be displayed using bar graphs. It comes from a presentation about an experimental study examining the mechanisms underlying elevated cytokine production seen in cirrhotic patients with infections; the speaker's comments are shown below.

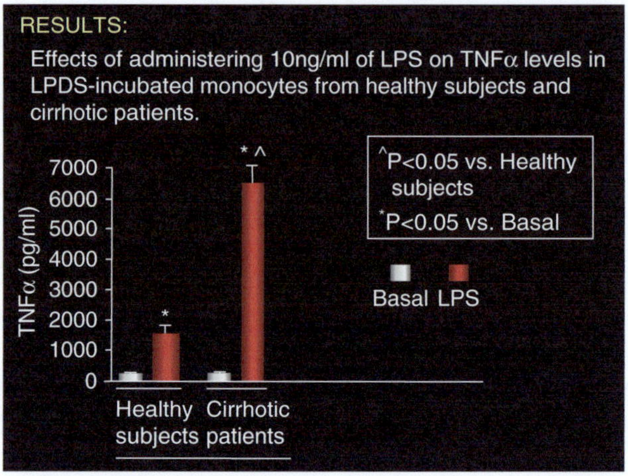

(Maria José Ramírez, PhD; used with permission)

This slide shows the effects of administering 10 ng/ml LPS on TNF-alpha levels in monocytes incubated with lipoprotein-deficient serum. The white bars show the baseline levels of TNF-alpha and the red bars show the TNF-alpha levels after the administration of LPS.

As you can see, there were no significant differences between the basal monocyte TNF-alpha levels in monocytes from healthy subjects and cirrhotic patients. Notice that administering LPS significantly raised TNF-alpha levels with respect to the baseline in both monocytes from healthy subjects and from cirrhotic patients. But the most important finding here is that the increase in TNF-alpha levels brought about by LPS administration was three times higher in monocytes from cirrhotic patients than in those from healthy subjects. This suggests that cirrhotic patients' monocytes are more sensitive than those of healthy subjects.

Emphasizing a Point

It is essential to emphasize the important points in your presentation. This can be accomplished in many ways: by repetition; by pausing; by modulating your voice, stance, and/or facial features; as well as by combining visual and oral information. However, you should never underestimate the power of telling your audience directly that a particular point is important. Just as physically pointing to the parts of a table or diagram help the audience to focus on what is important, comments like "*This is a crucial point*" or "*I can't overestimate the importance of this point*" will help drive home the important points in your presentation. Here are a few expressions that can help you emphasize a point in this way:

- *I want to stress that...*
- *I want to emphasize this point.*
- *This is a really important point.*
- *... is of paramount importance...*
- *This is the key to understanding...*
- *Whatever you do, do not forget that...*
- *Do not underestimate the role of...*
- *This is essential.*
- *This is important, so I want to make it crystal clear/I want to be sure that you get it.*
- *Let me point out that...*

Reiterating

- *In other words, ...*
- *What I am trying to say is that ....*
- *To put it another way, ...*
- *What this really means is that ....*
- *In a bit more detail,...*

Introducing Slides and Making Transitions

It is often more impressive to begin to introduce what comes next before you display the slide containing the information.

- *Now we're going to look at ....*
- *Again, ...*
- *This next slide shows ....*
- *And here you can see ....*
- *Let's look at a concrete example of this.*
- *This next case (image, graph, etc.) is an example of ....*
- *These data showed that ... is important, [advance to the next slide], but these other data show that ... is also important.*
- *Another factor that plays an important role in this process is ....*

It is important for your audience to know where they are at all times during your presentation. In addition to mapping out your presentation, it is important to remind them often where they are and where they are going to keep them from getting lost. It is equally important to make a smooth transition between points. Transition words can help fulfill both of these functions.

Certain words like *this*, *that*, *these*, *those*, and *the* point directly to something we mentioned early. If your previous slide displayed the results of your study, referring to "*these results*" when beginning to discuss the next slide makes a clear connection between the two. Another way to make this transition involves echoing parts of the material presented in the previous slide. For example, if one of the major findings presented on the previous slide was a difference in mortality rates between men and women, you might begin your discussion of the next slide with "*this difference between sexes*".

Transition words tell the audience whether the new information we are going to present will delve deeper into the material presented before, contrast with it, or leave it behind to move onto something completely different. The following words and expressions are useful for informing the audience about what kind of information will be presented next.

Adding new information along the same lines as that just presented: *what's more, moreover, further, furthermore, also, additionally, in addition, similarly, first, second, third, last, etc.*

Providing examples: *for example, for instance, in particular, let me give you a concrete example, let's have a look at an example from one of our patients*

Restating/reinforcing/explaining what has already been said: *again, in brief, in other words, in short, this means*

Introducing the logical conclusion of what we have shown: *therefore, thus, hence, so, consequently, in consequence, as a result, accordingly*

Summarizing: *in summary, to summarize, in conclusion, to conclude, in short*

One way of making the transition from one point to another is to ask a question. The audience will perk up their ears in anticipation of the answer. Here are some useful phrases to make the transition from the introduction, where you end by stating the hypothesis, objective, or research question, to the methods section, where you describe what you did to try to resolve the issue:

- *So, how did we go about investigating this?*
- *To test this hypothesis, we designed a series of experiments….*
- *To accomplish these objectives, we ….*
- *To investigate this question, we ….*
- *To determine whether A or B,*
- *To see whether this was true, we selected…*

To make the transition from the methods section to reporting your results, the following phrases can be useful:

- *So, what did we find?*
- *Let's have a look at what we found.*
- *We found some interesting results.*
- *The data we collected showed that our hypothesis was partially correct.*
- *Although the study is still underway, our preliminary results show….*
- *Our results are inconclusive...; nevertheless, it seems that….*
- *Contrary to what we expected, we found that ….*
- *Contrary to what you might expect….*
- *We were surprised to discover that ….*

To make the transition from reporting your results to discussing their implications or drawing conclusions, it is often convenient to summarize your results and say something like:

- *So, what does this tell us?*
- *All these results suggest that ….*
- *We can conclude that ….*
- *Taken together, these results point to three conclusions ….*
- *Although it is still too early to reach a definite conclusion, it seems that ….*
- *Although our data are limited, they do allow us to conclude that ….*
- *This made us wonder….*

Concluding Your Talk

The end of your talk is arguably the most important part. The audience's attention grows as they anticipate the end of a presentation, and you should not waste this opportunity to drive your point home. Be sure to state your main points clearly. Do not be afraid of repeating what you have said elsewhere in your talk.

Here are a few phrases that can help you:

- *I'd like to take a minute to go over these three take-home points.*
- *If you forget everything else I have discussed here today, remember these three points.*
- *That just about wraps things up.*
- *I hope you have enjoyed my talk.*
- *Thank you for your attention. I would be happy to try to answer any questions you might have.*

Dealing with Questions

Expressing incomprehension:

- *Could you be a bit more specific about…?*
- *Would you repeat the second part of your question?*
- *I'm afraid I still don't understand*
- *I'm sorry. I cannot understand your question. Could you please rephrase it and try to speak a bit more slowly?*

Stalling:

- *I am not sure I understood your question. Would you repeat it?*
- *I wonder if you could be a bit more specific about…*
- *What aspect of the problem are you referring to by saying…?*

Answering multiple questions:

- *There are two different questions here.*
- *With regard to your first question…*
- *Let me address your second question first.*

Disagreeing

- *With all due respect, I believe that there is no evidence of…*
- *I disagree with your comments on…*
- *I think that the importance of…cannot be denied*

Evading an issue

- *I'm afraid I'm not really in a position to be able to address your question yet.*
- *We'll come back to that in a minute, if you don't mind.*
- *I don't think we have enough time to discuss your comments in depth*
- *I would be happy to talk to you about this later.*
- *That certainly is an interesting question. Hugh McDonnell will probably be addressing it in his talk later in the session.*

We recommend that you practice some of these expressions and incorporate them into your presentation where appropriate. Increasing your "arsenal" of expressions will help to make you more confident and more fluent.

CHAPTER 4

Chapter 4 Common Mistakes in Language Usage

English is a difficult language both to speak and to write. In this chapter, we look at some common mistakes non-native-English-speaking physicians make in scientific presentations and give you some advice on how to avoid these pitfalls.

Mistakes in written English will leave a very bad impression on your audience. Whereas most attendees are likely to be tolerant of mistakes in spoken English, they are less likely to forgive mistakes in written English on slides. And it is only logical that audiences expect a higher standard in written material than in spoken material; after all, you will have had time to check and re-check your slides before you present them to the audience. Indeed, there can be no excuses for mistakes in spelling, grammar, and punctuation on slides. Whenever possible, you should ask a native English speaker to have a look at your slides – they might very well notice something you would never find on your own.

Here are a few common pitfalls in language usage:

1. Spelling mistakes

Spelling mistakes can be difficult to spot.

Incorrect: *Figure 1 shows a transrenal saggital scan of the right liver lobe.*
Correct: *Figure 1 shows a transrenal sagittal scan of the right liver lobe.*

Words with double consonants seem to be difficult for many people to get right. I wish I had a dollar for every time I have changed "saggital" to "sagittal"!

Always use your computer's spell checker; it works in PowerPoint as well as in Word (and in similar programs from other manufacturers). Set the language in Tools > Language and be sure to select the right variety of English from the list. However, unfortunately, you cannot count on the spell checker to catch all of your spelling mistakes for two reasons.

First, there are thousands of medical and technical words that your spell checker will not recognize. One way to deal with this problem is to painstakingly add medical words as the need arises to your personal dictionary (be sure to create separate dictionaries for different languages and for different varieties of English). You will have to double-check that all words are correct before you add them; if you make a mistake, you can edit the dictionary, but until you realize you have made a mistake you are at risk.

A second approach is to download special medical spell-checker software from the internet. There are many programs available: some are free and others

J. Giba and R. Ribes, *Preparing and Delivering Scientific Presentations*,
DOI: 10.1007/978-3-642-15889-6_4, © Springer-Verlag Berlin Heidelberg 2011

run the full spectrum from dirt cheap to outrageously expensive. Other websites offer "online spell checkers" in which you enter your text into a dialogue box. While these methods have definite advantages, they also have disadvantages.

For instance, most dictionaries include many words that, exist in English 6th are hardly ever used. Although the English language accepts many possible words, including many Latin constructs, we normally opt for a simpler, more common way of expressing the idea that comes from English's Germanic, Anglo-Saxon roots, when one exists. An example of this is the word "cardiopathies": this word can be found in all medical dictionaries and all English-speaking physicians would undoubtedly understand it; however when native-English-speaking physicians write and speak they invariably use "heart disease" for this concept.

We recommend creating your own personal dictionaries. You are unlikely to need the entire corpus of medical words found in even the simplest dictionaries, and you will benefit most from learning words as you need them. Researching the words yourself can help you become more familiar with them. You can optimize your time by looking for the spelling, meaning, and pronunciation all together.

Incorrect: *The mucus membrane lining the wall of the ureter….*
Correct: *The mucous membrane lining the wall of the ureter….*

Incorrect: *Asses the need for hemorrhoidectomy.*
Correct: *Assess the need for hemorrhoidectomy.*

Although spell checkers can be a great help, you cannot expect them to detect semantic errors. English has many words with nearly identical spellings that have different (sometimes very different!) meanings. *Mucus* is the noun and *mucous* is the adjective. *Asses* has two meanings (buttocks and donkeys), neither of which is appropriate in this context. Your computer's grammar check might vaguely warn you about possible errors if these words are different parts of speech (for example, a noun instead of a verb), but you shouldn't count on it. Some of these potential mistakes are mere typographical errors but others result from similarities between words that native speakers often confuse. Appendix 5 provides a short list of some common mistakes of this type, but you should know that there are many, many more.

Mistakes in grammar and punctuation are less likely than spelling mistakes, because slides tend to be telegraphic and have few complete sentences. However, some experts suggest that each slide should have at least one complete topic sentence that summarizes the purpose of the slide. Regardless of whether you include full sentences or just bulleted phrases, it is essential to pay careful attention to the basic principles of grammar, including word order, subject-verb agreement, countable-uncountable nouns, adverbs-adjectives, relative clauses, and prepositions.

2. **Noun strings.** Avoid long strings of nouns and adjectives. Three-word strings are difficult to decipher; longer strings can severely compromise clarity.

Unclear: *Transrectal ultrasound-guided prostrate biopsy false-negative prevention protocol*
Clear: *Protocol for the prevention of false-negative results in ultrasound-guided prostate biopsy*

Unclear: *High resolution CT scanner installation cost analysis*
Clear: *Analysis of the cost of installing a high resolution CT scanner*

Unclear: *Critical patient rights protection regulations consensus conference*
Clear: *Consensus conference on regulations to protect the rights of critical patients*

Expanding noun strings using prepositions can help to make your meaning clearer. If limited space on the slides requires you to use a few noun strings, be sure you at least give the expanded version in speech to make your meaning clearer.

3. **Word order.** Keep modifying words and phrases close to the words and phrases that they modify; otherwise, you may be asking for misinterpretation. This goes for prepositional phrases as well as participle clauses and applies to both spoken and written language. Here are just a few examples:

Ambiguous: "*Dr. Gottron saw a tapeworm working in Alabama.*" This says the tapeworm was working.
Clearer: "*When he was working in Alabama, Dr. Gottron saw a tapeworm.*"

Ambiguous: "*We found a spiculated lesion that was in the left breast on March 19.*" Presumably, if the lesion was in the breast on March 19, it was probably there on the days leading up to March 19, too.
Clearer: "*On March 19, we found a spiculated lesion in the left breast.*"

Ambiguous: "*Located near the aortic valve, we found an intimal tear in the ascending aorta.*" We were not located near the aortic valve.
Clearer: "*We found an intimal tear in the ascending aorta near the aortic valve.*"

Ambiguous: "*Many articles about inflammation have been written in the large bowel*".
Clearer: "*Many articles about inflammation in the large bowel have been written.*"

Incorrect: *We aimed to discover what are the mechanisms underlying this response.*
Correct: *We aimed to discover what the mechanisms underlying this response are.* Remember that in indirect questions the subject precedes the verb.

Incorrect: *I like very much elderly patients.*
Correct: *I like elderly patients very much.*

Do not place adverbs between verbs and their objects

4. **Subject-verb agreement.** Make sure that you use a singular verb with a singular or uncountable subject and a plural verb with a plural subject.

Incorrect: *The use of sterile gloves do not obviate the need for hand washing.*
Correct: *The use of sterile gloves does not obviate the need for hand washing.* Do not be confused by elements like preposition phrases that come between the subject and verb.

Incorrect: *The good new is that we have found a way around this problem.*
Correct: *The good news is that we have a way around this problem.* Remember that some uncountable nouns look plural but do not have a plural form. Other examples include *economics, glans, measles, mumps, physics, politics, statistics, rickets.*

Conversely, many nouns refer to a single object but take a plural verb, for example: *binoculars, forceps, goggles, pants, scissors, shears, spectacles, tongs, trousers.* These words are often used with *a pair of* and thus the entire expression takes a singular verb: *Forceps are used to extract the fragments* or *A pair of forceps is used to extract the fragments.*

Incorrect: *A number of samples was obtained.*
Correct: *A number of samples were obtained.* In this case, although it seems like the subject of these sentences, number, is singular in form, it is plural in meaning. We could replace with "A number of" with "Several" or "A few" or "Six". In any case, it always best to be as specific as possible.

Incorrect: *The data is confusing.*
Correct: *The data are confusing.* Many common Latin plurals are often wrongly treated as singular words. Other examples common examples include: *bacteria, criteria, ganglia, media, mitochondria, phenomena, and viscera.*

Incorrect: *Multiple metastasis were found in the lungs and liver.*
Correct: *Multiple metastases were found in the lungs and liver.* The rules for the formation of plurals in Latin and Greek medical words are summarized in Appendix 1.

5. British and American English

Incorrect: *Hemorrhage can lead to anaemia.*
Correct: *Hemorrhage can lead to anemia* or *Haemorrhage can lead to anaemia.*

Incorrect: *Traveler's diarrhoea can be caused by many different organisms.*
Correct: *Traveler's diarrhea* (or *traveller's diarrhoea*) *can be caused by many different organisms.*

Try to be consistent in all aspects of language. It may be difficult for non-native speakers to avoid mixing British and American vocabulary, pronunciation, and grammar; however, it is easy to be consistent in spelling, as differences in spelling tend to be systematic and easy to check automatically with your computer. Appendix 3 outlines the major differences between British and American spelling.

6. **Hyphenated adjectival expressions.** Remember that adjectives have no plural form in English; thus, hyphenated multiword expressions used as adjectives before nouns must be singular.

   Incorrect: *A 25-years-old woman presented with acute abdominal pain.*
   Correct: *A 25-year-old woman presented with acute abdominal pain.*

   Incorrect: *Patients underwent 15-minutes examinations.*
   Correct: *Patients underwent 15-minute examinations.*

   Incorrect: *They received a two-million-dollars grant for the study.*
   Correct: *They received a two-million-dollar grant for the study.*

7. **To: a preposition or part of the infinitive?**

   Incorrect: *I look forward to discuss this matter with you after the conference.*
   Correct: *I look forward to discussing this matter with you after the conference.*
   In this case, "to" is a preposition and the gerund is the only form of the verb that can be the object of a preposition.

   Incorrect: *Dr. Stroh will see to obtain the consent of the institutional review boards.*
   Correct: *Dr. Stroh will see to obtaining the consent of the institutional review boards.*

8. **Verb patterns.** When used in combination with other verbs, not all verbs are subject to the same grammatical treatment.

   Incorrect: *The patient refused signing the consent form.*
   Correct: *The patient refused to sign the consent form.*

Some verbs can only be followed by the full infinitive (with *to*). Appendix 3 lists other verbs like this.

   Incorrect: *We finished to collect the data and started to analyze them.*
   Correct: *We finished collecting the data and started to analyze them.*

Some verbs can only be followed by another verb if it is in the gerund form. Appendix 3 lists other verbs like this.

Some verbs can be followed by an infinitive or gerund with little or no difference in meaning. For example, both of the following sentences are correct and mean the same thing: *His blood pressure continued to drop* = *His blood pressure continued dropping*. Appendix 3 lists other verbs like this.

Some verbs can be followed by an infinitive or gerund but change their meaning depending on whether they are followed by an infinitive. For example, the meaning of stop in the following two sentences is very different: *Dr. Chan stopped to talk to the nurse* ≠ *She stopped talking to the nurse.* In the first example, Dr. Chan stopped another activity so that she could talk to the nurse. In the second, Dr. Chan terminated the activity of talking to the nurse. Appendix 3 lists other verbs like this and explains the difference in their meanings according to their usage.

Incorrect: *Our method allows to detect viable cells.*
Correct: *Our method allows us to detect viable tumor cells.*
Correct: *Our method allows viable tumor cells to be detected.*
Correct: *Our method allows detection of viable tumor cells.*

Some verbs can be followed by either an indirect object + a full infinitive or by a noun (or the gerund, which is the noun form of the verb). Appendix 3 lists other verbs like this.

Incorrect: *The results of the test enabled to reach the correct diagnosis.*
Incorrect: *The results of the test enabled reaching the correct diagnosis.*
Correct: *The results of the test enabled us to reach the correct diagnosis.*
Correct: *The results of the test enabled the correct diagnosis to be reached.*

Some verbs can be followed by an indirect object + a full infinitive but not by a noun or gerund. Appendix 3 lists other verbs like this.

Incorrect: *The medicine made him to vomit.*
Correct: *The medicine made him vomit.*

A few common verbs are followed by an object and the infinitive without *to*; these are discussed in Appendix 3.

9. Articles

Incorrect: *A MRI scan revealed a tear in the Achilles tendon.*
Correct: *An MRI scan revealed a tear in the Achilles tendon.*

Incorrect: *He completed an European certification program.*
Correct: *He completed a European certification program.*

Incorrect: *The operation took less than a hour.*
Correct: *The operation took less than an hour.* Remember that whether we use "a" or "an" depends on whether the noun following the indefinite article begins with a vowel sound or consonant sound (not whether it is spelled with an initial vowel or consonant).

Incorrect: *What a beautiful weather!*
Correct: *What beautiful weather!* We do not use indefinite articles with uncountable or plural nouns.

Incorrect: *The antibiotics are losing their effectiveness.*
Correct: *Antibiotics are losing their effectiveness.* We do not use the definite article when we refer to plural or uncountable nouns in the generic sense.

10. Abbreviations and acronyms

Space constrictions might tempt you to use abbreviations so that you can cram more information onto a slide. First, you should remember that your objective is not to cram as much as possible onto each slide – your objective is to make it as easy as possible for your audience to get your message.

In the right context, standard abbreviations can be very useful. You can safely use DNA for deoxyribonucleic acid in a presentation about genetic markers or MRI for magnetic resonance imaging in a presentation about neuroimaging; in these cases, it is not even necessary to define them the first time you use them. Likewise, you needn't worry about your slide looking like alphabet soup as long as you stick to standard abbreviations in a well-defined context (for example, DNA, RNA, mRNA, TNF, and PCR when talking to an audience of biomedical scientists familiar with genetics and inflammatory markers) or (CT, US, MRI, fMRI, PET, and PET-CT when talking to an audience of imaging specialists). The problem arises when the audience is unfamiliar with the topic of your presentation; in this case, you have to define and explain abbreviations just as you have to define any other term or concept that you cannot expect them to know beforehand.

Incorrect: *All patients underwent Magnetic Resonance Imaging (MRI).*
Correct: *All patients underwent magnetic resonance imaging (MRI).*

Incorrect: *Vaccination with Bacillus of Calmette and Guérin (BCG) can lead to a false-positive Mantoux test.*
Correct: *Vaccination with bacillus of Calmette and Guérin (BCG) can lead to a false-positive Mantoux test.*

When first introducing an acronym, do not capitalize the initial letters of the words represented, unless of course they would require capitalization in other contexts.

Be extremely careful about inventing your own abbreviations. Remember that, unlike in an article, where the reader can continually refer back to your definitions, the audience is usually unprepared to deal with more than one or two very obvious abbreviations. Thus, it is a good idea to repeat the definitions for these abbreviations on every slide and to say the full term while pointing to the abbreviations as often as possible.

Whatever you do, make sure that any abbreviation you decide to include cannot be confused with a standard abbreviation. For example, it wouldn't be a good idea to use TBI to represent "treated bacterial infection", because this abbreviation is widely used for "traumatic brain injury". You should also avoid forming acronyms with associations that might offend some members of the audience, like for example, referring to members of the percutaneous injection group as PIGs or to the superior occipital bone as the SOB.

Finally, be sure you know how to pronounce all the acronyms in your presentation. The pronunciation of acronyms is discussed in Chapter 10.

11. Relative clauses

There are two major kinds of relative clauses.

Non-defining relative clauses provide parenthetical information that does not change the essential meaning of the sentence.

Examples:

The patient, ***who is originally from Baltimore,*** *will be operated on tomorrow.*
Abdul Al-Dahar, ***whose family is extremely rich,*** *has written an excellent book on paraneoplastic skin disorders.*
Lyme disease, ***which is actually named after the town where it was originally described rather than after the person who first described it,*** *is caused by a spirochete transmitted by ticks.*
A team at the Hospital Clínic in Barcelona, ***where my sister works,*** *recently transplanted a trachea grown from the patient's own epithelial cells.*

Non-defining relative clauses are always separated from the rest of the sentence by commas; placing a comma before and after non-defining relative clauses reflects the brief pause in these positions in normal speech patterns. Be sure to pause in the correct place to avoid confusing your listeners.

Using "that" in non-defining clauses is a common mistake.

Defining relative clauses provide essential information that would change the meaning of the sentence if it was omitted.

Examples:

The patient ***that is in Box #9*** *(****=who is in box #9****) will be operated on tomorrow.*
Patients ***whose prostheses have loosened*** *are at increased risk for infection.*
Lyme disease is the most common serious disease ***that is transmitted by ticks = which is transmitted by ticks****.*
Nurses at the hospital ***where my sister works*** *are threatening to strike.*

Although it is not incorrect to use the relative pronouns "who" for people or "which" for things in defining clauses, it is possible to use "that" instead. Especially when your intonation is not perfect, using "that" saves your audience the trouble of wondering whether you intend the relative clause to be defining or non-defining.

12. Punctuation

Punctuation is necessary for clarity. Again, punctuation reflects the natural rhythm of speech, so it only goes to follow that you need to reflect punctuation in speech by pausing appropriately where periods (UK full stops), semicolons, and commas would appear in written language.

Furthermore, although your slides should have few complete sentences, that does not mean you can do away with punctuation all together. The following are some aspects of punctuation that are important to bear in mind when preparing slides:

Apostrophes: In contractions, apostrophes show that one or more letters have been admitted (can't = cannot), but it is uncommon to use this kind of contraction in scientific writing (although in speech, it usually sounds much more natural to use contractions than to omit them).

Incorrect: *CT: it's advantages and it's disadvantages*
Correct: *CT: its advantages and disadvantages*
Be sure not to write *it's* (=it is) for the possessive adjective *its*.

A much more common use of the apostrophe in scientific writing is the so-called Saxon genitive, which is used to show possession.

Incorrect: *All the dog's livers were cirrhotic.*
Correct: *All the dogs' livers were cirrhotic.*

We place the apostrophe before the S when we use it with singular nouns or with irregular plurals (*the patient's vital signs, the women's ages*) but after the S when we use it with regular plurals (*the patients' ages, the residents' lounge*). This usage is common in eponyms like *Crohn's disease, Behçet's syndrome, Ewing's sarcoma, Pacchioni's bodies, Raynaud's sign*, etc., although the *'s* is sometimes omitted (*Crohn disease*). As always, the most important thing is to be consistent. When more than one person is the possessor, the *'s* is applied only to the last person in the list (*Master and Johnson's results*) unless we want to refer to separate objects belonging to different persons (*Joel's, Jennifer's, and Michael's lab coats*).

Hyphens: Hyphens are often used to divide words at the end of a line of text. It can be complicated to divide words correctly and it does not look good on a slide, so we recommend that you avoid dividing words this way. However, if you must divide a word at the end of a line, check a dictionary to make sure you do it right.

Some compound words are always written with hyphens: The cardinal and ordinal numbers between 21 and 99 always need a hyphen between the tens and ones columns (*twenty-one, thirty-third*, etc.). Fractions are also normally written with a hyphen between the numerator and denominator (*two-thirds, three-fifths*), unless one of their components is written with a hyphen (*three twenty-fifths, twenty-one hundredths*). A few other common compounds are written with hyphens, for example: *Editor-in-chief, X-ray*, or *half-life*. Other compounds may be written as two separate words without hyphens, as a hyphenated compound, or as a single word without hyphens. Often all three ways of writing the compound coexist: in this case it is important to choose one approach and stick with it.

Incorrect: *Infection with Epstein Barr virus in young children is usually asymptomatic.*
Correct: *Infection with Epstein-Barr virus in young children is usually asymptomatic.*
Always use a hyphen to separate eponyms in syndromes, techniques, apparatuses, etc.: *Swan-Ganz catheter, Cheyne-Stokes respiration, Dandy-Walker syndrome, Guillain-Barré syndrome.*

Words made up by combining different elements from classical languages are normally written without hyphens in American English: *ventriculoperitoneal, temporomandibular, anteroposterior, hepaticoduodenostomy*, etc. It is also becoming increasingly common to write these words without hyphens in British English.

One of the areas in which non-native speakers often fail to hyphenate correctly is in temporary compound adjectival expressions. In this case, two or more words come together to form an adjective before a noun.

Incorrect: *She developed heparin induced thrombocytopenia.*
Correct: *She developed heparin-induced thrombocytopenia.*

There are an infinite number of possibilities; here are just a few examples: *a thirty-year-old woman, a two-million-dollar grant, cell-mediated immunity, a well-known author, well-defined margins,* etc.

Incorrect: *The author is well-known.*
Correct: *The author is well known.*

Remember that these expressions should not be hyphenated when they come after the noun. Likewise, ratios are hyphenated when named but not when explained: *the male-to-female ratio* but *the ratio of males to females.*

We do not use a hyphen with adverbs ending in –ly (*a recently reported case*) or with comparative or superlative adjectives (*a more effective treatment, a better designed protocol, the most probable explanation*). Moreover, we do not use hyphens with Latin expressions (*an in vivo experiment*), expressions with letters or numbers for classification (*grade A evidence, type II diabetes, neurofibromatosis type 2*), or nouns formed from phrasal verbs (*uptake, outcome, etc.*).

Incorrect: *CT revealed a semi-circular canal fistula.*
Correct: *CT revealed a semicircular canal fistula.*

Most prefixes are written without hyphens except when their use implies doubling a letter (*antibiotics* but *anti-inflammatory*, *intravenous* but *intra-arterial*, *preoperative* but *pre-existing*), although it is becoming more common to write even these words without hyphens (*intraabdominal*, *posttraumatic*, etc.). Again, the most important thing is to be consistent.

Incorrect: *Post Whipple procedure mortality is much higher in low-volume hospitals.*
Correct: *Post-Whipple procedure mortality is much higher in low-volume hospitals.*

However, prefixes should be hyphenated when used with proper nouns or numbers (*non-Hodgkin lymphoma, anti-Parkinson drug, pre-Billroth procedure*, etc.).

Capital letters: Capital letters are used for the first letter of the first word of every sentence. We also use capital letters for proper nouns and words derived from proper nouns: *Johannes Müller* and *Müllerian ducts. China* and *Chinese medicine.* Sometimes all the "important" words in a title are capitalized; in this case, nouns, pronouns, adjectives, verbs, adverbs, and numbers are considered "important": *Developing and Pilot Testing Quality Indicators in the Intensive Care Unit.*

Incorrect: *Dr. James Fischer will deliver the keynote address on monday, may 3.*
Correct: *Dr. James Fischer will deliver the keynote address on Monday, May 3.*

Remember that the names of the days of the week (*Monday, Tuesday*, etc.) and of the months (*January, February*, etc.) are capitalized.

Incorrect: *Patients' demographic data are shown in figure 1.*
Correct: *Patients' demographic data are shown in Figure 1.*

The words *appendix, figure*, and *table* are capitalized when accompanied by a number.

Incorrect: *The specimen was positive for helicobacter pylori.*
Correct: *The specimen was positive for Helicobacter pylori.*

We capitalize the genus in species names (*Staphylococcus aureus*) but not when we use them to refer to the group in general (staphylococci).

13. Verb tenses

Pay careful attention to the use of verb tenses. English verb conjugation is relatively simple, but the subtleties derived from the combination of time (past, present, future) and aspect (simple, continuous, or perfect) are often difficult for speakers of certain languages to grasp. Here are just a few examples of common mistakes:

Incorrect: *This graph is showing the differences between the two groups over time.*
Correct: *This graph shows the differences between the two groups over time.* The "showing" is not only in progress now – if you look at the same graph next week, it will show the same thing.

Incorrect: *We have done three experiments last year to test this hypothesis.*
Correct: *We did three experiments last year to test this hypothesis.* When we refer to an event that was completed at a specified time in the past, we use a simple tense.

Incorrect: *We studied 100 patients so far.*
Correct: *We have studied 100 patients so far.* When we want to talk about unfinished actions, we use a perfect tense.

If a member of the audience tells you "I would really appreciate a copy of your last slide", you should answer "*I'll send you one*" (and NOT "*I'm going to send you one*"). We reserve the use of "be going to" for future plans or predictions. When we make a spontaneous decision at the moment of speaking, we use "will".

Incorrect: *These pacemakers will be being manufactured in the People's Republic of China next year.*
Correct: *These pacemakers will be manufactured in the People's Republic of China next year.* We do not normally use the past perfect continuous, present perfect continuous, future continuous, or future perfect continuous tenses in the passive voice.

14. Unnatural passive constructions

Always use the active voice when it is more natural to do so. Do not be afraid of using personal pronouns like *I* or *we*.

Incorrect: *His gangrenous foot was amputated by somebody.* (Do not use an unspecified agent like somebody, them, or people in a byline).
Correct: *His gangrenous foot was amputated.*

Unnatural: *His gangrenous foot was amputated by a surgeon.* (By a surgeon is so obvious it is redundant. However, if the agent is surprising or unusual, it is natural to include it; for example, *His gangrenous foot was amputated by the chief of the tribe, who was also the medicine man* or *His foot was amputated by a madman).*
More natural: *The tumor was excised by Dr. Kaplan.*

Preferable in most contexts: *Dr. Kaplan excised the tumor.*

15. Numbers

Numbers pervade all aspects of life, and science as we know it would be impossible without them. Appendix 4 explains how to express numbers and mathematical expressions orally, as well as various aspects related to the use of numbers in scientific writing.

Incorrect: *Significance was set at* $p < 0,05$
Correct: *Significance was set at* $p < 0.05$

Remember that in English, we use periods not commas for decimal points. Thus, one million is written 1,000,000 (never 1.000.000) and five one-hundredths is expressed as decimal like this 0.05 (never 0,05).

Incorrect: *We conducted 50 15-minute interviews.*
Correct: *We conducted fifty 15-minute interviews* (or *50 fifteen-minute interviews*).
When two numbers are juxtaposed, write one of them as a word and the other as a numeral.

Final advice:
The mistakes you make will depend on your native tongue and your experience. Familiarize yourself with "false friends" and pay special attention to the differences between English and your native tongue. Most important of all, do not be afraid of making mistakes: remember that everybody makes mistakes; the important thing is to learn from your mistakes and avoid repeating them.

"The greatest mistake you can make in life is to be continually fearing you will make one."
Elbert Hubbard (1856–1915)

CHAPTER 5

Chapter 5 Delivering a Talk

Introduction

In this chapter, we discuss some important aspects of the actual delivery of a talk. First, we give some general pointers to help you improve the delivery of your presentation. Then we point out some specific aspects of delivering a presentation that are often overlooked by inexperienced speakers as well as some of the things that might go wrong and strategies for dealing with them. Finally, we give some general advice for controlling nervousness.

General Pointers

Familiarizing Yourself with the Room

Try to scout out the exact room where you are scheduled to give your presentation in advance. Pay attention to details that may be important like the position of the podium (if there is one), the seating arrangements, and the lighting. If at all possible, get up behind the podium and imagine yourself addressing the audience.

Arriving Early

Do everything in your power to arrive early for your presentation. Not only will this allow you to scout out the room as discussed above if you haven't had a chance to do so before, but it will also make it possible for you meet the chairman and other presenters if you haven't already. It is important to load your presentation in the computer and make sure it runs properly before your allotted slot. If the audience is relatively small and not extremely formal, you might find it helpful to introduce yourself to those who arrive early "to break the ice" and create a human connection that will help you to feel more comfortable during your talk.

J. Giba and R. Ribes, *Preparing and Delivering Scientific Presentations*,
DOI: 10.1007/978-3-642-15889-6_5, © Springer-Verlag Berlin Heidelberg 2011

Etiquette

After you are introduced to the audience, do not forget to direct your attention to the chairperson and briefly thank him or her (by name if possible) for introducing you. You should also let the audience know that you are happy to have the opportunity to address them and perhaps make a complimentary remark about their institution and city or country. Likewise, when you have finished your talk, it is polite to thank the audience for their attention. Finally, listeners might appreciate it if you offer them the opportunity to contact you in person or in writing after the talk.

Eye Contact

During the course of your talk, you should make eye contact with individuals in your audience. If you are speaking to a small group, you should try to engage every member of the audience at least a couple of times. If you are speaking to a larger group, you should try to engage a few individuals in every section of the audience at varying times during your talk. Do not follow the advice of those who urge you to fake eye contact by looking over the audience's heads and staring out into space – effective communication involves engaging the audience on a personal basis and eye contact is essential to human interaction.

Eye contact can provide you with important feedback from your audience. If they seem to be drifting off, pick up the pace a bit; if they seem to be lost and confused, slow down and repeat key points.

Modulating Your Speech

Change the speed and loudness of your delivery to emphasize key points. Speaking in a monotone will lull your audience to sleep. Audiences need to work hard to pay attention to the details of your talk – if you speak without inflection they will soon tire and lose interest. Slow down when you have to explain difficult or complex material. Pick up your pace when you notice the audience drifting off.

Controlling Your Movements

Body language can have a huge effect on audiences. Some studies have found that over half of interpersonal communication comes from body language and facial expression. You need to learn to be aware of your posture and movements. As mentioned in the previous chapter, videotaping yourself is an excellent way to learn how you use your body during a talk. You want to look natural and confident. Audiences tend to notice movements of your hands and feet, so you should avoid repetitive movements like tapping your foot or fingers, toying with your hair, tie, or belt, or jingling change or keys in your pocket (it's probably a good idea to empty your pockets before the presentation).

Your stance is also important. Do not slouch – stand up straight. Plant your feet firmly on the ground and do not rock back and forth. Do not grip the podium. When the arrangement of the arena allows it, it is often a good idea to move out from behind the podium, but if you do so, do not pace back and forth. Move closer to the audience to emphasize a point.

Do not turn your back on the audience to point out items on the screen. Make sure you have the proper means to point out anything you need to on the screen – your finger or a newspaper have obvious shortcomings. A long pointing rod was common in most lecture halls in the past; this allowed speakers to actually place the pointer against the projection screen to direct the audience's attention to a particular point while continuing to face the listeners. Nowadays, laser pointers are more common, but be careful – laser pointers magnify the tiny movements of trembling hands, so if you are nervous, be sure to anchor your arm against your side or on the podium. The computer's mouse pointer is a better way to point things out in a slide without turning your back on the audience; however, if you design your presentation using PowerPoint effects that let you make an arrow or circle appear on the screen with a click of the mouse or remote control, you won't need to worry about this at all.

Starting Out Strong

First impressions are lasting, so the beginning of a presentation is crucial. It is important to engage your audience early on, so do not make the fatal mistake of reading the first few slides to help you to feel confident – this mistake will inevitably have the opposite effect. The customary first slide containing the title of the presentation, your name, and a picture of your hospital should remain on the screen for at least the first 30 seconds of a 30-minutes presentation. During this short time, you must convince the audience that you are worth listening to and give them some idea what your paper is about. It will also give them time to adjust to your voice, accent, and way of talking. Many presenters make the mistake of rushing through the first slide so fast that the audience never even has a chance to read the title.

When you introduce the paper, you should provide a general overview of the topic you are going to discuss and try to make it attractive (see Chap. 3 for useful phrases for this purpose).

Making Smooth and Clear Transitions

Your audience should always know where they are in your presentation and where you are taking them. In addition to mapping the presentation early on, it is often a good idea to signal the start of each new section or change in topic with a dedicated slide. You should also signal this change with spoken language (see Chap. 3 for useful phrases) and voice modulation.

Accentuating Key Points

Given that most people remember only a small percentage of what they read and an even smaller percentage of what they hear, you would do well to make sure you're your audience both hears and sees your key points more than once. Make sure your key point is not buried in a crowded slide with other information that might distract your audience. You can draw attention to these key points by pausing and modulating your voice, but don't be afraid to tell your audience how important a key point is directly (see Chap. 3 for specific phrases). One very effective technique for accentuating key points is to pause both before and after delivering it.

Graphics

Do not neglect to explain the graphs and figures in your presentation clearly. No matter how simple and straightforward you think your graph is, remember that your audience is seeing it for the first time and they need time and help to digest it.

Finishing Strong

Audiences' attention rises to a peak in anticipation of the end of a talk, and a good speaker will take full advantage this fact. Be sure to leave plenty of time to present your conclusions, even if they are merely a rehashing of the main points you have made during the rest of the presentation. Many presenters do not reach the conclusion slide until they are running out of time; this means that they have to deliver the most important slide of the lecture in a rush, fighting against the customary warnings "Dr. Jones, you must finish your presentation" and finally cut their talk short after the final sentence "I'm sorry Dr. Jones, you are out of time".

Aim to reach your conclusion slide ahead of time and take your time to comment on each point carefully. In most cases, it is better to leave your conclusion slide projected after you have finished your talk, rather than replacing it with a blank screen or a cliché like "Thank You for Your Attention". Displaying your conclusions throughout the question and answer period gives the audience another opportunity to read your conclusions and think about your talk.

Finally, don't step down from the podium too soon – give the audience an opportunity to applaud after your presentation, and do not interrupt them. While you are receiving the applause, look at the attendees in the eye and use your body language (nodding slightly and smiling) to let the audience know that you are appreciating their response just as they have appreciated your lecture. After a brilliant presentation, both you and the audience deserve to enjoy the afterglow. Learn to read the audience's reactions by analyzing the applause for other presentations as well as for your own.

Feedback and Reflection

Feedback is the shortest word in the English language the contains all of the letters A,B,C,D, E, and F. The audience's opinion is paramount. While a good lecturer will have paid enough attention to the audience during the presentation to gauge their response to his or her talk, the more feedback you have the better. If you know someone in the audience or in the organization, ask them for constructive criticism about your talk. However, be aware that out of politeness many people will simply tell you that you did fine to avoid hurting your feelings. Additionally, a few days after your talk, you should ask for the attendees' evaluation forms to get an idea of how the audience evaluated your performance and the lectures given by the rest of the faculty.

After your talk, perhaps on your way back home, reflect on what went well and what went badly in your presentation. Make a few notes about your impressions to help you in future presentations. Every talk can be improved, but we tend to reflect only on presentations that did not go as well as we had hoped. This is a mistake – dissecting and analyzing in depth only your worst presentations will undermine your confidence. Think about what went right as well as about what went wrong; both types of information can help you become a better speaker.

Perseverance

Immediately after they step down from the podium to sit down in the safe anonymity of their seats in the first row of the conference room, many non-native English speakers breathe a sigh of relief and consider that their careers as presenters at international congresses has come to an end. This is, in our opinion, one of the most common and erroneous reactions to the stress involved in the delivery of your first lecture in English.

Only by persevering with a second presentation can you discover the amazing differences between your first presentation and consecutive ones. It is a pity to waste all the effort devoted to the first presentation and throw it away because you will probably see that most negative aspects of delivering a scientific paper will be absent in successive presentations.

Dealing with Nervousness

If you are not at least slightly nervous before an important presentation, you are probably not human. Mark Twain got it right when he said "There are two types of speakers: those that are nervous and those that are liars". It is perfectly natural to feel nervous before a talk. There are many ways to deal with pre-talk nervousness; however, nothing will serve you better than the knowledge that you have prepared thoroughly for the occasion. If you have prepared thoroughly for your talk, you should be eagerly awaiting the opportunity to present your material to your audience. Like an athlete before a match or a performing artist before a recital, you will be "keyed up" and may experience the feeling of "butterflies in your stomach". You should be confident and know on some level that you have no need to be nervous. The important thing is not to let your nervousness overwhelm you and to use your excitation to your advantage. The best speakers, like the best athletes and performers, are able to channel their nervous energy. If you can manage to do this, your enthusiasm about your topic is sure to be transmitted to your listeners and the success of your presentation is almost guaranteed.

Everybody has their own way of dealing with nervousness. Many claim that the power of positive thinking works wonders. I always find it helpful to imagine myself giving a great performance before I go up on stage. It also gives me great comfort to think that the organizers, the audience, and my colleagues – nearly everyone has a stake in my talk – and they all want me to succeed. Remind yourself that you can do this and you can do it well.

Once you have successfully dealt with your nervousness in one talk, you will learn that physiological nervousness is a natural process and will naturally run its course. Typically, your nervousness will disappear as you get into your material. In future talks, dealing with nervousness will not present such a big problem, and after you gain considerable experience, you may even come to enjoy the restless anticipation of your talk.

Dealing with Technical Problems during the Presentation

Technical problems are among the most stressing situations a lecturer may face, and you would be well advised to develop strategies for dealing with them. While these problems are often unavoidable and unpredictable, the important thing is for you to maintain control of the situation. The number of things that have a certain probability of going wrong is limited, and you can prepare to deal with the most common ones. Knowing and rehearsing a few key words and phrases can give you the confidence you need to handle the vast majority of technical problems that might come up during your talk.

In English, we use the interjection "Oops" to acknowledge a minor mistake. Try to get used to saying "oops" whenever something happens unexpectedly. Here we show you how to use this interjection naturally in the context of some of the most common problems.

Imagine that you are behind the podium and the microphone is not working properly. In the best of possible worlds, a helpful technician will always be nearby standing ready to intervene and fix the problem; however, unfortunately, in this imperfect world we live in, you may have to say something like:

"Oops! I am afraid that my microphone is not working properly. May I use yours, please?"

Or maybe you go to point something out on the screen and discover that your laser pointer is not working properly. You do not want to wait until you are in the middle of your presentation to find out that you don't know how to say "laser pointer" in English!

To ask for one, you can say something like:

"Oops! I've noticed that my pointer is not working properly. May I have another one, please?"

If the projector is not working properly, you may have to ask for help:

"*Oops! There seems to be a problem with the projector.*" A comment like this is usually enough to bring a technician running. Look at the chairperson and ask for help directly: "*Can I get some help with this, please?*"

If the computer is not working properly, you will have to ask for help:

"Oops! I am afraid that there is a problem with computer. Can somebody help me with this or can I have another one, please?"

Presentations by surgeons, interventional and cardiac radiologists, and endoscopists, among other specialists, often have embedded videos. Videos can present a wide variety of technical problems due to incompatibilities between different software and hardware and it can be a real nightmare when they don't work properly. I have seen many potentially excellent lectures spoiled by technical problems with videos. It is essential to make sure that your videos run properly. Many people make the mistake of checking the videos on a different computer from the one you are going to use during the presentation. Videos must be double-checked on the very computer you are going to use for the talk.

No matter what precautions you take, there is a chance that one of your videos will not run properly. If that happens, don't stop the presentation to try to solve this problem because you run the risk of losing your momentum and the audience will disconnect immediately. Unless you are capable of summarizing the contents and message of the video and have rehearsed doing so, it is best just to say something like: "*Oops! This video is not running properly. I'll show you the next slide*" and move on to the next slide immediately without making any other comments about the video the audience is missing.

If your USB memory fails, you should have another one at hand. As mentioned in the last chapter, it is also a good idea to have your presentation on another medium, like a CD or DVD in case the problem is with the connection rather than with your memory stick.

If the electricity goes off and the room is dark, you should say something like: "*Oops! As you can see, or indeed do not see at all, the lights have gone out.*" On occasions like this, the chairperson should step in and will probably adjourn the session until the technical problem can be fixed.

Dealing with Disruptions

While you cannot be held responsible for disruptions that are beyond your control, you are responsible for how you handle disruptions, so you should do everything in your power to take control of the situation and minimize the disturbance. If there is a lot of noise coming from somewhere outside the lecture hall, for example, you should ask yourself whether closing the door or window will mitigate the noise. In an informal setting, it might be easiest for you to close it yourself, but usually you can ask something like "*Can somebody please close that door?*" and say "*thank you*" when they have done so. If a loud train drowns you out, stop talking until it passes and the audience can hear you.

Rarely, you will come up against disruptive members of the audience. These are usually easy to recognize: they tend to be seated in the last rows of the conference room and speak to each other during your talk. This will hardly ever be a single individual; rather, there will usually be a small group. The best way to engage them is to look them in the eye from the beginning of the presentation. Most inexperienced lecturers make the mistake of looking away from these attendees because they undermine their self-confidence. Do not make this mistake. If absolutely necessary, you can stop your presentation and address them with an extremely polite question in the softest tone you can muster:

"*I am sorry; do you have any questions or comments?*"

Once you have interrupted the delivery of your talk and have made him/her the absolute- and negative- protagonist of the floor, he/she will surely stop disturbing you and the audience.

CHAPTER 6

Chapter 6 The Dreaded Questions and Comments Section

Introduction

Many physicians would feel confident enough to give an oral communication at an international congress if they could be sure that they would not have to take any questions from the audience afterward. This is perfectly logical if we consider that you can at least prepare for if not control the vast majority of things might happen during your presentation, but there is a lot more uncertainty involved in taking questions from the floor.

Nevertheless, we will see that there is much you can do to prepare for the dreaded question and comments section. Having the tools you need to deal with this situation will help to take the dread out of this section until you are more fluent and more experienced in speaking in English and can look forward to fielding questions as a way to explain your research in greater detail.

General Tips for Dealing with the Question and Answer Section

1. Don't wait for the chair to ask the audience whether they have any questions – ask them directly yourself. "*I would be happy to try to answer any questions you might have.*" "*Please feel free to ask me any questions. I will do my best to answer them.*" "*What questions do you have?*" or simply "*Any questions*?" If nobody asks a question after a minute or so, you might formulate one yourself.
2. Repeat each question out loud to make sure that you have understood it correctly and that everybody in the audience knows what you've been asked. If you are unsure, rephrase the question or ask your interlocutor to clarify the question.
3. Show respect for your questioner and his or her question. Don't rush to give an answer. Always wait for your interlocutor to finish, unless he or she is rambling on so much you are worried that you won't have enough time to answer and take more questions. Most of our readers would be happy to let the clock tick away, but you WILL reach a point where you look forward to the opportunity of explaining your work in greater detail. When you reach

J. Giba and R. Ribes, *Preparing and Delivering Scientific Presentations*,
DOI: 10.1007/978-3-642-15889-6_6, © Springer-Verlag Berlin Heidelberg 2011

this point, in cases like this you will have to be tactful. First raise your hand (most people will respond to this gesture) and then interrupt your speaker saying something like "*So, if I understand correctly, you are asking …?*"

4. Give yourself time to think about your answer, and if you are sure you have understood the question correctly but are unable to answer it, be honest. "*That's a very good question, but I'm afraid I don't have the answer to it right now. That data won't be available for several months*". If it is some part of your research that another team member is responsible for "*That's Mary Genet's field – talk to me later and I'll tell you how to contact her*". If you consider that the interlocutor has asked something that you should know, offer to research the answer and report back to him or her later (and then follow through on your promise!). If it is a more general question about your topic, suggest resources that would help your audience tackle the question on their own or ask whether anyone else in the audience knows the answer. Don't ever be afraid to say "*I'm sorry, I don't know*" – remember that nobody has all the answers all the time. People will appreciate your sincerity.
5. If somebody asks you about something that you explicitly covered in your talk, answer just the same. Maybe you didn't really explain it so clearly during the presentation. However, if somebody repeats a question that's already been asked, it's better to say something like "*I think we've already covered that*" unless you think that you might not have answered the question sufficiently the first time; in that case, you should try to answer the question using a different approach. If somebody asks a completely irrelevant question, say something like "*That's not really part of my topic today.*"
6. Avoid arguments. You have everything to lose and nothing to gain. Don't lose your composure. Don't show irritation or anger. If you feel attacked, don't get on the defensive – practice restating hostile, negative questions in a positive way and answering them. If you can't do that, simply say "*this is not the time or place for this debate. I'll be happy to discuss this with you later.*"
7. When you finish answering the question, look at your interlocutor and ask "*Does that answer your question?*" or "*I hope that helps to answer your question*". If your interlocutor says you haven't really answered his or her question, ask for clarification or offer to discuss it in greater detail afterward.
8. Prepare a short statement to end your question and answer session, thanking your audience for their interest. If there seem to be a lot of unanswered questions when you run out of time, apologize for not being able to deal with them and offer to make yourself available later.

However, the above suggestions may not be so easy to carry out if your English is not up to par and you have little or no experience speaking in public. The following anecdote illustrates some of the feelings many non-native English-speaking physicians might go through in their first presentations in English:

After a short free communication on the MRI follow-up of the Ross operation (the surgical replacement of a patient's aortic valve by the pulmonary valve and the replacement of the latter by a homograft) that had gone reasonably well considering my inexperience, I was waiting nervously for the round of questions that would inevitably follow my presentation.

An English radiologist asked me a question I could barely understand. When I asked him politely to repeat his question, he obediently repeated his question – word for word and speaking just as quickly as he had the first time! Seeing that I was still unable to understand the question, the chairman roughly translated it into more international and easily understandable English, and I answered it as best I could. This was the only question I was asked since we had run out of time.

And the worst was still to come. I wasted the whole morning recreating the scene over and over. "How could I not have understood such an easy question? How could I have spoiled so many hours of research and study? I even thought that people recognized me as "the one who didn't understand a simple query …"

Let us analyze what happened to me and see if we can get anything useful from it.

1. Understanding the question was the greatest challenge.
2. Native English speakers tend to be more difficult to understand.
3. The interlocutor did exactly what I asked him to do.
4. A good chairperson will always help you.
5. Time is limited and you can take advantage of this fact.
6. If you perform less than optimally, put it behind you and move on. Nothing is easy the first time – the important thing is to learn from your mistakes.

These points lead to some recommendations:

1. Answering the question was not especially difficult for me. I knew my material very well; after all, I had done the research on which the presentation was based and deeply pondered what to include and what leave out. The main difficulty I had was understanding what the English radiologist had asked me.
2. When a non-native English speaker asks you a question, you are talking to someone like you who has spent many hours struggling to learn to speak English. This person may speak better or worse than you, but he or she will almost certainly be easier to understand than a native speaker and is likely to have more empathy for any difficulties you might have in understanding the question.

   On the other hand, when you have to deal with a native English speaker there are two main types of interlocutors:
   Type A is a colleague who reduces his normal rhythm of speech to make it easier for you to understand the question.
   Type B is a colleague who does not make any allowance for the difference between native and non-native English lecturers. Needless to say, I faced a type B interlocutor in my first international presentation.

3. I was unlucky, but I exacerbated the situation by clumsily asking him to repeat his question. As beginners, we often pretend to know more than we actually do, so I asked him to repeat his question and that is what he did. If I had been modest enough to admit that I couldn't understand his question because my

English was not so good and requested my interlocutor to ask his question again more slowly and in a different way so I could understand it, he would have been morally obliged to do so.

"I'm afraid I don't understand your question. Would you please reformulate it in a different way, please?"
"Could you please speak more slowly?"

4. Look to the chairperson to help you. If you still cannot understand the question after asking the questioner to reformulate it, you can look at the chairperson and say. "*I'm afraid I still cannot understand the question. Can you help me with this?*"
5. The question and answer period is, at worst, a few minutes of stress. Do not let such a short period of time keep you from a potentially successful career in international medicine.
6. If you cannot understand or answer a question, few people will remember it. Don't be so hard on yourself--think for a moment how you performed the first time you did anything in your life, for example, the first time you picked up a tennis racquet or a golf club. Skills develop with experience.

When deciding on a strategy to deal with questions, it is also important to consider the different types of interlocutors you may face:

- *Type 1*: The interlocutor who wants to know a particular detail of your presentation. These are easy to handle by just answering his questions.

  "What diameters do you measure in the aortic root?"
  "Annulus, Valsalva sinuses, and sinotubular junction."

You can prepare for many of the questions that are likely to come up. You should know your material better than anybody, so you can most likely anticipate specific questions. A trick some presenters use is to intentionally leave out an important point – they know that somebody is likely to ask them about that point and have their answer prepared. This technique allows you to control what you want to talk about.

- *Type 2*: The interlocutor who wants the audience to notice his sound knowledge of the subject being discussed. These are the easiest of all to handle since they do not formulate questions as such but make a point of their own. In this case, the speaker's replies tend to be shorter than the interlocutor's questions or comments and the clock continues to run, leaving no room for another dreaded question. You only need to say something like:

  "Thank you for adding that interesting point"
  "I do agree with your comments."
  "We are planning to include this point in our next paper on …"

- *Type 3*: This is an interlocutor who strongly disagrees with your points. This is obviously the most difficult for a beginner to handle. The most important piece of advice we can give you here is to be humble when you try to defend your points and be very careful not to challenge your interlocutor. A few useful approaches include:

"I will consider your suggestions on ..."
"This is a work in progress and we will consider including your suggestions ..."

Unbelievably Ineffective Approaches

We have witnessed several variations on an ineffective approach to the questions and comments section by presenters with poor English. The following "humorous" description of different ways of dealing with the awful situation of not knowing what to say before a crowded audience is not our own invention. We have personally witnessed all these different cases on many occasions. At first it is difficult to believe that anybody might adopt such an approach, but we assure you that these attitudes are displayed so often that you can almost get used to them.

Typically, after struggling through a presentation, the presenter looks at the chairman, who says:

"*Any questions or comments from the audience*?"

A member of the audience asks simple, straightforward question like:

"*Did you set any cut off measurements to select the patients with adenomyosis in your study?*"

And the presenter either remains completely mute for 30 seconds or combines the mute approach with saying something totally inappropriate like "*Thank you very much*" or "*Sorry, I don't speak English*"!!!

The chairman, who cannot believe what is going on, finally says:

"*We must proceed to the next presentation*".

This is obviously not a very effective way of dealing with questions, yet the presenters who adopt this approach do survive and can proudly include the successful presentation in their curriculum vitae. However, there are better ways.

Stalling

When you are not confident enough in your capability to cope with the dreaded questions and comments section, one way to approach this terrifying situation is to lengthen your talk a bit so that you shorten the dreaded section that comes after our presentation. Plan your talk to be a little longer than the time allocated - when the chairman stops you, "accept" just one question from the audience. You can act like a basketball team in the lead at the end of a game: by stalling. The idea is to try to keep control of the ball – or in this case, the word – as long as possible and when your time of possession is about to end, shoot for the hoop – give a concrete answer. This stalling technique should only be considered as a last resort.

In this case, you might plan to go over the time limit for your presentation by a few seconds, leading to a situation like the following one:

- Chairman:

  "Dr. Tachinardi you must keep an eye on the time."
- Speaker:

  "I'm about to conclude, Mr. Chairman."
- Chairman:

  "I'm afraid your time is over. Any questions or comments from the audience?" (10 s)
  So now the remaining time is 90 – 10 = 80 **seconds**
- Doctor from the audience:
  - *"Did you set any cut off measurements to select the patients with adenomyosis in your study?"* (15 s)
  - With *65 seconds remaining,* the speaker counters with
  - *"Would you please repeat your question, I'm not sure I've caught it entirely?* (10 s)

  So now the remaining time is 55 seconds
- Doctor from the audience:
  - *"Did you set any cut off measurements to select the patients with adenomyosis in your study?"* (15 s).

  40 seconds left
- Speaker:
  - *"OK, I see, you mean that if we have set any cut off measurement in order to select the patients with adenomyosis in our study"* (16 s).

And then the speaker, *with only 24 seconds left*, will address the *ONLY* question of the section.

Shooting from the Hip

The expression "to shoot from the hip" means to go with your instincts and respond to a situation, in this case a question from the audience, without stopping to analyze it and to weigh possible options. This technique is useful for speakers whose poor comprehension skills may make it impossible for them to understand the details of the question. The speaker prepares general answers to questions containing certain key words. Hearing the key word in a question triggers the prepared response. For example, on hearing the word "angiogenesis" in a question, the speaker responds with a short prepared statement about the role of angiogenesis in the subject under discussion or in his or her work. In this way, the speaker is able to deliver a relatively relevant response without fully understanding the question.

The Guardian Angel

This technique requires the collaboration of a friend or colleague whose comprehension of spoken English is quite good. The team prepares answers to several likely questions and assigns a number code to each. The "guardian angel" sits in the first rows of the audience, and informs the speaker which answer would be most appropriate to answer the question by holding up a finger. Although the success of this technique ultimately depends on the team's ability to anticipate the questions that will be asked, just having a "guardian angel" in the audience can do wonders to booster the speaker's confidence and alleviate the loneliness of the first-time presenter.

Chapter 3 lists some phrases that may help you to deal with the dreaded questions and comments sections, regardless of the approach you decide to adopt.

CHAPTER 7

Chapter 7 Chairing a Session

If you have been invited to chair a session or moderate a roundtable discussion at an international meeting, the odds are that your medical English will probably be quite good. These opportunities usually do not come your way until after you have submitted many papers and given many presentations.

Chairing a session is a great responsibility and is not to be taken lightly. The chairperson or moderator should oversee all the activity in the conference room. Chairpersons need to do their best to make sure that everything goes as smoothly as possible, and a good chairperson will help the audience and speakers to connect with one another.

Chairpersons and moderators do not usually need to prepare a presentation of their own, but they are expected to review all the recently published material on the subject under discussion as well as all the abstracts of the talks to be presented in the session. Chairpersons should also prepare questions for each presentation in case the audience has no questions or comments.

Finally, a chairperson has to think about all the possible things that might go wrong and develop strategies to deal with them. There are tricky or even embarrassing situations that can be difficult for even an experienced moderator to handle.

In the following sections, we provide some useful language to help you carry out a chairperson's duties. Mastering certain key expressions will provide you with a sense of fluency, and as we have said repeatedly, fluency builds confidence and the lack of fluency undermines confidence. Knowing these key sentences and using them appropriately can help make chairing a session easy and enjoyable. On the other hand, if you do not know these expressions, you run the risk of getting tangled up in an embarrassing situation. If you are a first-time chairperson, rehearsing some of these sentences will help you feel confident. If you have a chaired a session before, it might be a good idea to review them. There is nothing like being prepared.

The organization will often provide you with some guidelines. These can range from very general recommendations like "Be strict with time control" to very specific like "Clearly state the room number and title of the session in your opening statement". It goes without saying that you should read the guidelines very carefully and question the organizers about any points you are unsure of.

A good moderator can really help ensure the success of a session. A bad moderator can really make things more difficult for the people he's there to help. Prepare your role and make sure that your contribution is a positive one.

J. Giba and R. Ribes, *Preparing and Delivering Scientific Presentations*,
DOI: 10.1007/978-3-642-15889-6_7,

Opening the Session

You should open the session with about 30 seconds of well-planned comments. As moderator, you can set the tone for the entire session. Try to communicate enthusiasm for the topic.

The following comments can be useful for introducing the session:

- *Good morning* (before 12 p.m.) *ladies and gentlemen. My name is Sarah Han and I want to welcome you all to this workshop on hernia repair. My co-chair is Dr. David Wilson from Shouldice Hospital in Toronto.*
- *Good afternoon* (after 12 p.m. until dark – sessions do not usually go on into the evening). *My name is Bill Harmon and this is session 43 on noninvasive mechanical ventilation. Please take a seat and disconnect your cell phones. We have a lot of exciting material to cover in a short time. We will listen to ten six-minute lectures with a two-minute period for questions and comments after each. Afterwards, provided we are still on time, we will have a final round of questions and comments from the audience, speakers, and panelists.*
- *Good morning. We will proceed with the session on mitral valve replacement. We have an exciting lineup of speakers. However, as many papers have to be delivered, I encourage the speakers to keep an eye on the time.*
- *Good morning. My name is Judith Carbonell and I'm going to moderate this session on the minimally invasive approach to Charcot foot. We have an interesting lineup of speakers, and I'm sure you're all looking forward to hearing what they have to say. However, there are a few points we need to cover before we get started....*

Introducing Speakers

Be sure to arrive in time to meet the speakers before the session. Introduce yourself to them and do your best to learn how to pronounce their names. Ask them whether they have any questions about how the session will be run and whether there is any particular question they would like you to ask them after their presentation. Try to create a relaxed atmosphere to dispel nervousness and put everybody at ease, perhaps lightening the mood with a joke.

It is essential for you to know how to pronounce all the words in the titles of the talks you will introduce. This should be part of your "homework" in preparing your role as moderator long before you ever arrive at the congress venue, but this is a good time for you have a final practice run.

When time allows, you should introduce each speaker with a few relevant points of information, for example: universities attended, past positions or past activities, current position, current activities and focus, etc. Try to learn this information and to introduce each speaker as if you knew him or her well – even if you have never met them before.

The following comments can be useful for introducing speakers:

- *Our first speaker is Dr. Chandragupta from Xanit International Hospital in Málaga, Spain, who will present the paper: "Stent grafting of abdominal aneurysms".*
- *Please welcome our first speaker, Dr. Joseph Byrd, from the University of Iowa Medical Center. Dr. Byrd's paper is entitled "Can closer follow up improve cure rates for sequential therapy?"*
- *We'll be getting underway with a talk by one of our field's most renowned specialists, Dr. Maria Bruni, from Milan University. Dr. Bruni trained at Stanford and Berkeley in the States and everybody is sure to be familiar with her work on brainstem lesions at Carlisle Center in London. She holds the Medici Chair of Neuroscience at the University of Milan and is the coordinator of the NOW trial. Today, Dr. Bruni will be presenting the paper "We've come a long way baby – where we stand and where we're heading".*

The following speakers are introduced in nearly the same way with comments like:

- *Our next lecturer is Dr. Mark Adams. Dr. Adams comes from Brigham and Women's Hospital, Harvard Medical School, and his presentation is entitled "Non-operative treatment of intraosseous ganglion".*
- *Next is Dr. Diane Shaw from Beth Israel Deaconess Hospital, presenting "Stem cells in hepatic surgery".*
- *Dr. Lars Olsen from UCSF is the next and last speaker. His presentation is: "Metastatic disease. Pathways to the heart."*

Once the speakers finish their presentation, the chairperson is supposed to say something like:

- *Thank you Dr. Vida for your excellent presentation. Any questions or comments?*

The chairperson often comments on presentations, although this is not strictly necessary:

- *Thank you Dr. Vida for your presentation. Are there any questions or comments from the audience?*
- *Thanks, Dr. Shaw. That was a very comprehensive presentation. Does the audience have any comments?*
- *Thank you very much for your clear presentation on this always controversial topic. I would like to ask a question. May I?* (Although being the chairperson you are the one who gives permission, it is common courtesy to ask the speaker for permission.)
- *I'd like to thank you for this outstanding talk Dr. Olsen. Any questions?*
- *Thanks a lot for your talk Dr. Ho. I wonder if the audience has any questions?*

Adjourning

The following comments can be useful for adjourning the session:

- *I think we all are a bit tired, so we'll have a short break.*
- *The session is adjourned until 4 p.m.*
- *We'll take a short break. Please do not go far – the session will resume in 15 minutes.*
- *We'll take a 30-minute break. Please fill out the evaluation forms.*
- *The session is adjourned until tomorrow morning. Enjoy your stay in Vienna.*

Concluding the Session

The following comments can be useful for finishing the session:

- *I'd like to thank all the speakers for your interesting presentations and the audience for your comments. (I'll) see you all at the congress dinner and awards ceremony.*
- *The session is over. I want to thank all the participants for their contribution. (I'll) see you tomorrow morning. Remember to pick up your attendance certificates if you have not already done so.*
- *We should finish up here. We have another group coming in. I look forward to discussing some of these topics with you later on.*
- *That just about wraps everything up. Thank you all for your participation.*
- *I'm afraid we have run out of time. It has been a pleasure to share this session with you. I have learned a lot and I am more motivated than ever to learn more about this fascinating subject. I look forward to the publications that will undoubtedly result from the studies you have underway.*

Managing the Questions and Answers Session

The chairperson is responsible for creating a productive atmosphere and ensuring the smooth running of the session. To do this, a chairperson will probably need to ask a few questions, especially at the beginning of the session when the audience does not usually make any comments at all. Getting things started and warming up the session are among the chairperson's duties, and if nobody in the audience comes forward with a question, the chairperson should step in and ask something.

Begin by calling for questions:

- *Are there any questions?*
- *Does anyone have a question for Dr. Chandler?*
- *Well, now that we've heard from all the speakers, I'd like to open the session to the floor.*

If nobody raises their hand or speaks up:

- *Well, I have two questions for Dr. Adams. Do you think surgery alone is the mainstay of prostate cancer treatment or do you see a role for hybrid interventions with radiotherapy? And second: What, in your opinion, should the role of chemotherapy be in this surgical algorithm?*

Once the session has been warmed-up, the chairperson should only ask questions or add comments to manage the timing of the session; so, if the session is behind schedule (as it usually is), the chairperson is not required to participate unless strictly necessary.

The chairperson does not have to demonstrate to the audience his or her knowledge about the topics discussed by asking too many questions or making comments. The chairperson's knowledge of the subject is not in question.

Managing the Time

Being the timekeeper is one of the chairperson's most important roles. Each speaker has a certain amount of time for his or her presentation. For the session to work, it is essential to keep to schedule. A few words of warning in private before the session often helps to let speakers know you are serious about this role. Do not let speakers go over time, because this would steal time from the questions/comments time and from later speakers. Do not be intimidated. Chairpersons should be forthright in telling speakers to stay within the time limit:

- *Dr. Berlusconi, your time is almost over. You have 30 seconds to finish your presentation.*
- *Dr. Ho, you are running out of time.*
- *Dr. Russell, we're going over time. Please finish up.*

If the speaker does not finish his presentation on time, the chairperson may say:

- *Dr. Berlusconi, I'm sorry but your time is over. We must proceed to the next presentation. Any questions, comments?*
- *We're out of time, Dr. Ho. We need to move on to the questions.*
- *Dr. Russell, I'm afraid I'm going to have to ask you to stop talking. Your time is up.*

After introducing the next speaker, the following comments will help you handle the session:

- *Dr. Goyen, please keep an eye on the time, we are behind schedule.*
- *We are running behind schedule, so I remind all speakers you have six minutes to deliver your presentation.*

Although it is unusual, sometimes there is some extra time and this is a good chance to ask the panelists a general question about their experience at their respective institutions:

- *As we are a little bit ahead of schedule, I encourage the panelists and the audience to ask questions and offer comments.*
- *I have a question for the panelists: What percentage of the total number of aortic operations is performed on children at your institution?*

Technical Problems

A wide variety of technical problems can occur. The chairperson is responsible for seeing that they are taken care of with as little disturbance to the proceedings as possible.

As a chairperson, you are supposed to fill in the gaps in the session; so, if a technical problem occurs, you need to keep the audience engaged and focused on the session in the meantime. One good approach is to start a discussion about the current situation regarding the topic of the session in the panelists' and audience's institutions or countries.

- *Regarding training in laparoscopic surgery, how are things going in Italy, Dr. Carini?*
- *As for as the use of ventricular assist devices, what's the deal in Japan, Dr. Hashimoto?*
- *How is the current situation in Germany regarding repayment policies?*
- *May I ask how many cerebellar tumor resections you are performing yearly at your respective institutions?*
- *Dr. Olsen, can you tell us how you deal with this problem at Johns Hopkins?*

By opening a discussion on how things are going in different countries, the not-too-fluent chairperson shares the burden of filling in the gaps with the panelists. This trick rarely fails and once the technical problem is fixed the session can go on normally with nobody in the audience noticing the chairperson's lack of fluency.

The following comments can be useful for dealing with some of the most common problems that might come up:

Computer Not Working:

- *I am afraid there is a technical problem with the computer. In the meantime, I would like to take this opportunity to comment about …*
- *The computer is not working properly. Until it is running again, I encourage the panelists to offer their comments about the presentations we have already seen.*
- *It seems the computer is on the blink. The hotel staff have informed us that we should have a new one up and running within a quarter of an hour. I propose that we take our break now rather than at 11:30.*

Lights Gone Out

- *The lights have gone out. We'll take what will hopefully be a short break until they are repaired.*

- *As you see, or indeed do not see at all, the lights have gone out. The hotel staff have told us it is going to be a matter of minutes, so do not go too far; we'll resume as soon as possible.*

Sound Gone Off

- *Dr. Hoffman, we cannot hear you. There must be a problem with your microphone.*
- *Perhaps you could try this microphone?*
- *Please, would you use the microphone? The rows at the back cannot hear you.*
- *Can somebody please help Dr. Lin with her microphone. It doesn't seem to be working properly.*

Lecturer Lacks Confidence

It can be painful to watch another human being suffering under the stress of giving a presentation. Apart from trying to create a supportive environment and put presenters at their ease from the very beginning, there is often little a chairperson can do to help a nervous speaker.

If the lecturer is speaking too quietly:

- *Dr. Cavett, would you please speak up? It is difficult to hear you.*
- *Dr. Oroaco, please speak up a bit. The people at the back cannot hear you.*

If the lecturer is so nervous he/she cannot go on delivering the presentation:

- *Dr. Olsen, take your time. We can proceed to the next presentation, so whenever you feel OK and ready to deliver yours, it will be a pleasure to listen to it.*

Finally, a chairperson will have many occasions to make specific comments and questions, which obviously depend on the content of the presentations themselves. This type of comment is usually relatively easy for a well-versed chairperson to make. Here are just a few examples:

- *Dr. Alley, I'm not sure which lesion you are talking about. Would you please use the pointer to show us?*
- *Dr. Wilson, would you please point out the number of patients excluded from the study?*
- *Dr. Negroponte, did you perform an abdominal ultrasound or a CT scan on this patient?*
- *Dr. Maier, did you perform the brain MRI on an emergent basis?*
- *Have you had any adverse anaphylactic reactions to this type of contrast material?*
- *Dr. Olsen, I can't see the lesion you are talking about. Can you point it out?*
- *Do you use 12F catheters for this purpose?*
- *Dr. Pons, I'm afraid that the video is not running properly. Could you try to fix it so we can see your excellent slides of MR images?*
- *Dr. Hashimoto, why didn't you use a 0.0035 stiff guidewire to cross the stenosis?*
- *Dr. Soares, are you currently using the harmonic scalpel in cases like this one?*
- *Dr. Mas, is trackability that important in these cases?*
- *Do you do enhanced lumbosacral spine MRIs in all postoperative patients?*
- *Do you perform preprocedural pelvic MRI in all patients undergoing pelvic tumor resections?*

In conclusion, if you are honored with an invitation to chair or moderate a session at a congress, do your best to ensure the best experience for both the audience and the speakers. Prepare your role thoroughly – be prepared to deal with all the expected (and unexpected!) situations that will arise. Fulfilling your responsibilities can be a rewarding and enriching experience.

CHAPTER 8

Chapter 8 Attending an International Scientific Congress

This chapter provides you with tips and language tools that will be useful in your itinerary to an international course: airport, plane, customs, taxi, hotel check-in, and finally, the course itself. We recommend upper-intermediate English speakers to go over this information quickly. Intermediate English speakers, however, should review this section thoroughly. Unless you have overcome the conversational hurdles in the scenarios that come before the course, you will not feel like delivering your presentation and you might not even be able to arrive at the course venue!

Don't let your lack of fluency in day-to-day English undermine your ability to deliver a good or even great presentation. Colloquial English and formal medical English are two different domains, and it is possible to give a great presentation without being fluent in other situations. However, a sound knowledge of colloquial English can help you to reach the podium without suffering a nervous breakdown.

International congresses are an excellent opportunity to improve your skills as a presenter as well as your skills in both medical and colloquial English. You can learn a lot from watching other speakers. Pay attention to how each speaker presents his material. Note down aspects that you liked and didn't like in each presentation. Think how the speaker could have done better. You might want to try out different styles to help you find your own unique style, and congresses are full of different models to choose from.

Most beginners do not go alone to their first courses abroad. While they are usually relieved to be accompanied because this means they do not have to cope with the language difficulties on their own, going with colleagues has the unfortunate result that most non-native-English-speaking physicians return to their respective countries without having uttered a single word in English. Although you may consider it unnatural to speak English with your colleagues, it is often the only opportunity you will have to speak English during the course, since over 90% of your conversations will be with your fellow countrymen. In parties of more than two people, it's virtually impossible to do this simple exercise.

Traveling alone is the only way to be sure that you will speak English during an international course, and this may be the only opportunity for non-native-English-speaking physicians to keep their English alive throughout the year. Do not waste this excellent opportunity to practice both colloquial English and medical English.

The following anecdote illustrates the level of uncertainty young non-native-English-speaking physicians face when they attend their first international

J. Giba and R. Ribes, *Preparing and Delivering Scientific Presentations*,
DOI: 10.1007/978-3-642-15889-6_8, © Springer-Verlag Berlin Heidelberg 2011

meetings. It was Ramon's first European Congress of Radiology in Vienna. When he was waiting to be attended to at the registration desk and given his congress bag, somebody asked him: "Have you got your badge?" Not knowing what badge meant he said "no" since he was unlikely to have something on him he did not even know the name of. The person who had asked the question commanded: "Go to that line," so he obediently got into the line without having any idea what he was waiting for. Such experiences can be disheartening: when you are supposed to be an expert who is arriving to give a lecture and something like this happens to you, all you want to do is go back home.

The material that follows is in no way exhaustive, but familiarity with the exchanges below will help get you through the vast majority of situations you are likely to come up against in your itinerary.

Travel and Hotel Arrangements

Airport

Getting to the Airport

- *How can I get to the airport?*
- *How long before departure do we need to be at the airport?*

Checking In

- May I have your passport and flight tickets, please?
  Of course, here you are.
- Are you Mr. Vida?
  Yes, I am.
  How do you spell it?
  V-I-D-A

IMPORTANT: rehearse the spelling of your name, since you will probably be asked to spell it many times.

- Here is your boarding card. Your flight leaves from gate 43.
  Thank you.
- You are only allowed two carry-on items. You'll have to check that larger bag.

Questions a Passenger Might Ask

- *I want to fly to London leaving this afternoon. Is there a direct flight? Is it via Zurich?*
- *Is it direct?*
  Yes, it is direct/No, it has one stopover.

- *Is there a stopover?*
  Yes, you have a stopover in Berlin.
- *How long is the stopover?*
  About 1 h.
- *Do I have to change planes?*
  Yes, you have to change planes at …
- *How much carry-on luggage am I allowed?*
- *What is the weight limit?*
- *My luggage is over the weight limit. How much is the excess baggage fee?*
- *Is a meal served onboard?*
  Yes, lunch will be served during the flight./No, just a snack.
- *What time does the plane to Chicago leave?*
- *When does the next flight to Chicago leave?*
- *Can I get onto the next flight?*
- *Can I change my flight schedule?*
- *What's the departure time?*
- *Is the plane on time?*
- *When does the plane get to San Francisco?*
- *Will I be able to make my connection?*
- *I have lost my hand luggage. Where is the lost property office?*
- *How much is it to upgrade this ticket to first class?*
- *I want to change the return flight date from Boston to Madrid to November 30th.*
- *Is it possible to purchase an open ticket?*
- *I have missed my flight to New York. When does the next flight leave, please?*
- *Can I use the ticket I have or do I need to pay for a new one?*

Announcing Changes in an Airline Flight

- Our flight to Madrid has been cancelled because of snow.
- Our flight to Chicago has been delayed; however, all connecting flights can be made
- Flight number 112 to Paris has been cancelled.
- Flight number 1145 has been moved to gate B12.
- Passengers for flight number 112 to London go to gate 7. *Hurry up! Our flight has been called over the loudspeaker.*

At the Boarding Gate

- We will begin boarding soon. Please have your boarding card and picture ID ready.
- We are now boarding passengers in rows 24 through 36.
- May I see your boarding card?

Arrival

- *Pick up your luggage at the terminal.*
- *Where can I find a luggage cart?*

- *Where is the taxi rank?*
- *Where is the subway stop?*
- *Where is the way out?*

Complaining About Lost or Damaged Luggage

- *My luggage is missing.*
- *One of my bags seems to be missing.*
- *My luggage is damaged.*
- *One of my suitcases has been lost.*

Exchange Office

- *Where is the exchange office?*
- *What is the rate for the dollar?*
- *Could you change 1,000 euros into dollars?*

Customs and Immigration Control

- May I see your passport, please?
- Do you have your visa?
- What is your nationality?
- What is the purpose of your journey?
  I am attending a medical conference/I am speaking at the RSNA/I am giving a series of lectures at the University of Kentucky/I will be doing a fellowship at Johns Hopkins …
- How long do you plan on staying?
- Empty your pockets and put your wallet, keys, mobile phone, and coins on this tray.
- Remove any metallic objects you are carrying and put them on this tray.
- Open your laptop.
- Take off your shoes. Put them in this tray too.
- Do you have anything to declare?
  No, I don't have anything to declare / No, I only have personal effects / Yes, I am a doctor and I'm carrying some surgical instruments / Yes, I have bought six bottles of whisky and four cartons of cigarettes in the duty free shop.
- How much currency are you bringing into the country?
  I haven't got any foreign currency.
- Open your bag, please.
- I need to examine the contents of your bag.
- May I close my bag?
  Sure
- Please place your suitcases on the table.
- What do you have in these parcels?
  Some presents for my wife and kids.

- *How much duty do I have to pay?*
- *Where is the exchange office?*

During the Flight

Very few exchanges are likely during a normal flight. If you are familiar with them, you will realize how fluency influences your mood positively. Conversely, if you need a pillow and are not able to ask for it, your self-confidence will shrink, your neck will hurt, and you will not ask for anything else during the flight. On my first flight to the States I did not know how to ask for a pillow and tried to convince myself that I did not actually need one. When I looked it up in my phrasebook, asked for it, and the stewardess brought the pillow, I was glad I had bothered and managed to fall asleep.

Do not let lack of fluency spoil an otherwise perfect flight.

- *Is there an aisle/window seat free? (I asked for one at the check-in and they told me I should ask on board just in case there had been a cancellation).*
- *Excuse me, you are in my seat.* Oh! Sorry, I didn't realize.
- Fasten your seat belt, please.
- Your life-jacket is under your seat.
- Smoking is not allowed during the flight.
- *Please would you bring me a blanket/pillow?*
- *Is there a business class seat free?*
- *Can I upgrade to first class on board?*
- Would you like a cup of coffee/tea/a glass of soda? *A glass of soda, please.*
- What would you prefer, chicken or beef/fish or meat? *Beef/Fish, please.*
- *Is there a vegetarian menu?*
- *Stewardess, I'm feeling bad. Do you have anything for flight-sickness? Could you bring me another sick-bag, please.*
- *Stewardess, I have a headache. Do you have an aspirin?*
- *Stewardess, this gentleman is disturbing me.*

In the Taxi (*US* Cab)

Think for a moment about taking a taxi in your city. How many sentences do you suppose would be exchanged in normal, and even extraordinary, conditions? We assure you that with fewer than two dozen sentences you will solve more than 90% of possible situations.

Asking Where to Get a Taxi

- *Where is the nearest taxi rank?*
- *Where can I get a taxi?*

Basic Instructions

- *Hi, take me downtown/to the Sheraton hotel, please.*
- *Please would you take me to the airport?*
- It is rush hour; I don't go to the airport.
- Sorry, I am not on duty.
- It will cost you double fare to leave the city.
- *I need to go to the Convention Center.*
- Which way do you want me to take you, via Fifth or Seventh Avenue? Either one would be OK.
- *Is there a surcharge to the airport?*
- *Can you please help me with my luggage?*

Concerning Speed in a Taxi

- *To downtown as quick as you can.*
- Are you in a hurry? *Yes, I'm in a hurry.*
- *I'm late; please hurry.*
- *Slow down!*
- *Do you have to drive so fast? There is no need to hurry. I am not in a rush at all.*

Concerning Smoking in a Taxi

- *Would you mind putting your cigarette out?*
- *Would you mind not smoking, please?*

Asking to Stop and Wait

- *Stop at number 112, please.*
- Which side of the street?
- Do you want me to drop you at the door?
- *Pull over; I'll be back in a minute.*
- *Please, wait here a minute.*
- *Stop here.*

Concerning the Temperature in a Taxi

- *Would you please wind your window up? It's a bit cold.*
- *Could you turn the heat up/down/on/off?*
- *Could you turn the air conditioning on/off?*
- *Is the air conditioning/heating on?*

Payment

- *How much is it?*
- *How much do I owe you?*

- *Is the tip included?*
- *Do you have change for a twenty/fifty (dollar bill)?*
  Sorry, I don't (have any change).
- *Keep the change.*
- *Would you give me a receipt?*
- *I need a receipt, please.*
- *I think that is too expensive.*
- *They have never charged me this before. Give me a receipt, please. I think I'll make a complaint.*
- *Can I pay by credit card?*
  Sure, swipe your card here.

At the Hotel

Checking In

- May I help you?
  Hello, I have reserved a room under the name of Dr. Viamonte.
- For how many people?
  Two, my wife and me.
- *Do you need my ID?*
- *Do you need my credit card?*
- How long will you be staying? *We are staying for a week.*
- You will have to wait until your room is ready.
- Here is your key.
- Enjoy your stay. *Thank you.*
- *Is there anybody who can help me with my bags?*
- Do you need a bellboy? *Yes, please.*
- I'll have someone bring your luggage up.

Preferences

- *Can you double-check that we have a double room with a view of the beach/city …?*
- *I would like a room at the front/at the rear.*
- *I would like the quietest room you have.*
- *I would like a non-smoking room.*
- *I would like a suite.*
- How many beds? *I want a double bed/a single bed.*
- *I asked for two single beds.*
- *I'd like a king-sized bed.*
- *I'd like a queen-sized bed.*
- *We will need a crib for the baby.*
- *Are all of your rooms en suite?* Yes, all of our rooms have a bath or shower.
- *Is breakfast included?*

- *Does the hotel have a parking lot* (UK: car park)?
- *Is there a parking garage nearby?*

The Stay

- *Can you give me a wake-up call at seven each morning?*
- *There is no hot water. Would you please send someone to fix it?*
- *The TV is not working properly. Would you please send someone to fix it?*
- *The bathtub has no plug. Would you please send someone up with one?*
- *The people in the room next to mine are making a racket. Would you please tell them to keep it down?*
- *I want to change my room. It's too noisy.*
- *What time does breakfast start?*
- *How can I get to the city center?*
- *Can we change euros into dollars?*
- *Could you recommend a good restaurant near the hotel?*
- *Could you recommend a good restaurant?*
- *Would you give me the number for room service?*
- *I will have a cheese omelet, a ham sandwich and an orange juice.*
- *Are there vending machines available?*
- *Do you have a fax?*
- *Do you serve meals?*
- *Is there a pool/restaurant …?*
- *How do I get room service?*
- *Is there wireless/internet connection?*
- *The sink is clogged.*
- *The toilet is running.*
- *The toilet is leaking.*
- *My toilet overflowed!*
- *The toilet doesn't flush.*
- *The bath is leaking.*
- *My bathroom is flooded.*
- *The bath faucets (UK* taps) *drip day and night.*
- *The water is rust-colored.*
- *The pipes are always banging.*
- *The water is too hot.*
- *The water is never hot enough.*
- *I don't have any hot water.*

Checking Out

- *How much is it?*
- *Do you accept credit cards?*
- *Can I pay in dollars/euros?*

- *I'd like a receipt, please.*
- *What time is checkout? Checkout is at 11 a.m.*
- *I would like to check out.*
- *Is there a penalty for late checkout?*
- *Please would you have my luggage brought down?*
- *Would you please call me a taxi?*
- *How far is the nearest bus stop/subway station?*

Complaints

- *Excuse me, there is a mistake on the receipt:*
- *I have had only one breakfast.*
- *I thought breakfast was included.*
- *I have been in a single room.*
- *Have you got a complaints book?*
- *Please would you give me my car keys?*
- *Is there anybody here who can help me with my luggage?*

Course Example

General Information

By way of example let's review some general information concerning a course program, focusing on terms that beginners may not know.

Dress Code

Formal dress is required for the Opening Ceremony and for the Social Dinner. Casual wear is acceptable for all other events and occasions (although formal dress is customary for lecturers).

Commercial Exhibition

Participants will have the opportunity to visit representatives from pharmaceutical, diagnostic and equipment companies, and publishers at their stands to discuss new developments and receive up-to-date product information.

Although most beginners don't talk to salespeople due to their lack of fluency in English, **talking to salespeople in commercial stands is a good way to practice medical English and, by the same token, receive up-to-date information on equipment and devices you currently use, or will use in the future, at your institution.**

Disclosure Statements

To avoid commercial bias, speakers have to report whether they have significant relationships with industry or not.

As far as commercial relationships with industry are concerned, there are three types of speakers:

1. Speakers (spouses/partners, and planners) who have no reported significant relationships with industry.
2. Speakers who have reported receiving something of "value" from a company whose product is related to the content of their presentations.
3. Speakers who have not provided information about their relationship with industry.

Faculty

Name and current posts of the speakers:

- Russel J. Curtin, MD. Staff Surgeon. Division of Bariatric Surgery, Beth Israel Deaconess Medical Center, Boston, MA

Guest Faculty
Names and current posts of speakers coming from institutions other than those organizing the course:

- Fergus B Schwartz, Professor of Radiology and Otolaryngology, Head and Neck Surgery, New York School of Medicine; New York University Medical Center, New York, NY

Congress venue (where the course is to be held, e.g., hotel, university, convention center …):

Continental Hotel
32 Park Street, 23089 …
Phone: … /Fax: …
E-mail: continentalhotel@hhs.com

To reach the venue from the city centre (Charles Square) take the U1 underground line (green). Leave the train at Park Street and take the exit marked Continental Hotel. Traveling time: approximately 10 min.

Registration

Generally, you will have been registered beforehand and you will not have to register at the course's registration counter. If you do have to register at the congress venue, the following are some of the most usual exchanges that may take place during registration:

Surgeon:	*May I have a registration form, please?*
Course attendant:	Do you want me to fill it out (*UK* fill it in) for you?
	Are you a surgeon?
	Are you an ESS member?
	Are you attending the full course?
Surgical resident:	*No. I'm a resident (nurse/technologist)*
Course attendant:	Can I see your chairman's confirmation letter?
Surgical resident:	*I was told it was faxed last week. Would you check that, please?*
Surgeon:	*I'll pay by cash/credit card.*
	Charge it to my credit card.
	Would you make out an invoice?
	Can I have a receipt, please?
Course attendant:	Do you need an invoice?
	Do you want me to draw up an invoice?
Surgeon:	*Where should I get my badge?*
Course attendant:	Join that line.

Course Planning

The basic idea whenever you attend an international course is that you must rehearse those situations that are inevitably going to happen beforehand. By doing this, you will keep embarrassing situations that can catch you off-guard to a minimum. If only Ramon had rehearsed (at home!) the meaning of the word "badge", he wouldn't have been caught by surprise on his first course abroad. **You only need to know a few words, set phrases, and collocations to be fluent in a course environment, and we can assure you that knowing them will give you the confidence you need to make your participation in the course a personal success.**

The first piece of advice is: read the program of the course thoroughly and ask your more experienced colleagues about the words and concepts you don't know or look them up in the dictionary. Since the program is available before the course starts, go over it at home; you shouldn't have to read the scientific program at the course's venue.

"Adjourn" is one of those typical program terms with which one gets familiar once the session is "adjourned". Although many could think that most terms are going to be integrated and understood by their context, our intention is to go over those "insignificant" terms that may prevent you from optimizing your time at the course.

The following terms often appear in course plans:

- *Satellite symposia (singular: symposium)*: Scientific events sponsored by pharmaceutical firms where new drugs, techniques, or devices are presented to the medical community.
- *Plenary sessions*: These events take place usually at midday; they gather all participants around outstanding members of the medical community.
- *Cases of the day*: A number of cases covering different sections of your specialty. Participants can submit their diagnosis.
- *Categorical courses*: An important subject is discussed focusing on the needs of general physicians.
- *Refresher courses*: A concrete topic is reviewed in depth by experts in that particular field.
- "*... meets*" *sessions*: The purpose of these sessions is to forge closer ties between some invited countries and the congress. There are dedicated sessions for the specialist communities of these nations to demonstrate the excellence of specialists in their countries to congress attendees.
- *Special focus session*: The aim of a special focus session is to deal with a relevant "hot topic", presented in such a way as to promote debate between the panelists and the audience.
- *Scientific session*: The Scientific Committee selects, from all the abstracts submitted, the most outstanding basic and clinical research work, and invites the authors to make a presentation of their methods and conclusions (usually not longer than 10–15 min). A round of questions and/or comments is usually permitted.
- *Adjourn:* Close (break or recess) at the end of a session.

CHAPTER 9

Chapter 9 Conversation Survival Guide

Introduction

Many professionals who can hold their own fairly well in a technical conversation in English find themselves at a total loss for words in other situations. This is only to be expected. Work is a big part of life, taking up a large proportion of your waking life. Dedicated scientists and physicians often spend an even larger proportion of their time immerged in their work than their counterparts in other professions, and the chances are that most of your contact with the English language is related to your work. That is well and good, but work is not everything. When you travel outside your own country, you will need to use English in many different situations. Remember – fluency builds self-confidence and the lack of fluency breeds insecurity.

This chapter does not pretend to replace phrasebooks; on the contrary, we encourage you to use phrasebooks, no matter how good your English is. It can be a great help to have the translation of common expressions and the specific vocabulary you might need in a specific context.

Many upper-intermediate speakers do not take a phrasebook when traveling abroad. Convinced that they do not need help constructing basic sentences, they would be ashamed to be seen reading one. This is a mistake. You needn't be ashamed of your interest in improving your English, and phrasebooks can be used in different ways. As your English improves, you will find that you no longer need to read the translations, except for a few words, but you will still need help finding natural ways of saying things.

For example, whenever you go out to eat, you should review the key words and useful expressions from your phrasebook. It will not take more than a few minutes, and you will enjoy your meal much more if you have ordered it with unbelievable fluency and precision. Whereas it is recommendable for upper-intermediate speakers to review your phrasebook before going to a restaurant, it is absolutely essential for lower-intermediate speakers. Lower-intermediate speakers need to review and rehearse the expressions they will need to be able to order what they want to eat.

Phrasebooks can also be extremely helpful in unfamiliar environments. It is virtually impossible to be fluent in all situations, even in your own language. Think, for example, of a visit to a florist. How many names of flowers do you know in your own language? Probably fewer than a dozen. What about a trip to the

J. Giba and R. Ribes, *Preparing and Delivering Scientific Presentations*,
DOI: 10.1007/978-3-642-15889-6_9, © Springer-Verlag Berlin Heidelberg 2011

jeweler's? Do you know enough vocabulary related precious metals, gemstones, and designs to help you get a special gift for a special friend? Every conversation scenario requires its own specific vocabulary and a phrasebook can give you the help that you need to be fluent in many situations. So, do not be ashamed of carrying and reading a phrasebook; it may be the quickest way to gain fluency in unfamiliar scenarios that can try our English and our self-confidence in English.

This conversation guide is different from a phrasebook in that it does not include translations. Its purpose is to provide a "survival guide", a basic tool for upper-intermediate speakers who are actually perfectly capable of understanding all the usual exchanges, but might have some difficulty in finding natural ways to express themselves in certain less common scenarios. Less fluent speakers can benefit greatly from deciphering the meaning of the phrases below – use a dictionary or ask a friend for help if necessary.

Greetings

- *Hi.*
- *Hello.*
- *Good morning.*
- *Good afternoon.*
- *Good evening.*
- *How are you?* Fine, thank you.
- *How are you getting on?* All right, thank you.
- *I am glad to see you.*
- *Nice to see you (again).*
- *How do you feel today?*
- *How is your family?*
- *Good bye.*
- *Bye.*
- *See you later.*
- *See you soon.*
- *See you tomorrow.*
- *Give my regards to everybody.*
- *Give my love to your children.*
- *Good night.*
- *Have a good trip.*
- *I hope to see you again next year.*
- *I'm looking forward to seeing you at the European Congress in May.*

Introductions

- *This is Mr./Mrs/ Miss/ Ms . ...*(Use Mrs. /'mɪsɪz/ for married women, Miss /mɪs/ for single women, and Ms. /mɪz/ when you do not know if a woman is married or not. Ms. is always correct and can avoid embarrassment.)

- *These are Mister and Missus …*
- *My name is …*
- *What is your name?* My name is …
- *Pleased/Nice to meet you.*
- *Let me introduce you to …*
- *I'd like to introduce you to …*
- *Do you know Bill Atkins?*
- *I'd like you to meet Bob Jarvis, chief of vascular surgery at Memorial.*
- *Have you already met Mr. …? Yes, I have.*

Personal Information

- *What is your name?* My name is …
- *What is your surname/family name/last name?* My surname/family name/last name is …**Remember that you should practice spelling your name.**
- *Where are you from?* I am from …
- *Where do you live?* I live in …
- *What is your address?* My address is …
- *What is your email address?* My email address is …(the symbol @ is read "at" in English)
- *What is your phone number?* My phone number is …**Say phone numbers digit by digit to make sure you are understood**
- *What is your mobile phone/cellular number?* My mobile phone/cellular number is …
- *How old are you?* I am 43/ I will be 38 years old in May…
- *Where were you born?* I was born in …
- *What do you do?* I am a vascular surgeon./ I am a general practitioner./ I am a pediatric neurologist
- *What do you do?* I specialize in laparoscopic knee surgery / I have a private practice in a small town / I am on the staff at Reading Memorial …

Expressions of Courtesy

- *Thank you very much.* You are welcome/ Don't mention it.
- *Would you please …?* Sure, it is a pleasure.
- *Excuse me.*
- *Pardon.*
- *Sorry.*
- *Cheers!*
- *Congratulations!*
- *Good luck!*
- *It doesn't matter!*

- *May I help you?*
- *Here you are.*
- *You are very kind.*
- *That is very kind of you.*
- *Don't worry, that's not what I wanted.*
- *Sorry to bother/trouble you.*
- *Don't worry!*
- *What can I do for you?*
- *How can I help you?*
- *Would you like something to drink?*
- *Can I get you something to drink?*
- *Would you like a cigarette?*
- *I would like …*
- *I beg your pardon.*
- *Have a nice day.*

Speaking Languages

- *Do you speak English/Spanish/French …?* I do not speak English/Only a bit/ Not a word.
- – *Do you understand me?* Yes, I do. / No, I don't.
- *Sorry, I do not understand you.*
- *Could you speak slowly, please?*
- *How do you write that?*
- *Could you write that down?*
- *How do you spell that?*
- *How do you pronounce that?*
- *Sorry, what did you say?*
- *Sorry, my English is not very good.*
- *Sorry, I didn't get that.*
- *Could you please repeat that?*
- *I can't hear you.*

At the Restaurant

Do you know what the most common dish ordered by non-fluent speakers around the world? It is not a hamburger or hotdog or spaghetti, but a varied concoction called "the same for me". To avoid having to communicate with the waiter, many non-fluent English speakers automatically link their fate to whoever is sitting next to them.

A quick look at a phrasebook a few minutes before dinner will provide you with enough vocabulary to ask for whatever you want. Seize the opportunity to have a short exchange with the waiter or waitress, and do not miss the chance to eat what you want just because you are unsure of your English and have not taken a few minutes to prepare for the experience.

Preliminary Exchanges

- *Hello, have you got a table for three people?*
- *May I book a table for a party of seven at 6 o'clock?*
- What time are you coming, sir?
- *Where can we sit?*
- *Is this chair free?*
- *Is this table taken?*
- I'll bring you the menus in a moment. Would you like something to drink in the meantime?
- Are you ready to order?
- *Waiter/waitress, I would like to order, please.*
- *Could I see the menu, please?*
- *Could you bring the menu, please?*
- *Can I have the wine list, please?*
- *Could you give us a table next to the window?*
- *Could you give me a table on the mezzanine?*
- *Could you give us a table near the stage?*

Ordering

- *We'd like to order now.*
- *Could you bring us some bread, please?*
- *We'd like to have something to drink.*
- Here you are.
- *Could you recommend a local wine?*
- *Could you recommend one of your specialties?*
- *Could you suggest something special?*
- *What are the ingredients of this dish?*
- *I'll have a steamed lobster, please.*
- How would you like your meat, sir?
  Rare/medium-rare/medium/well-done.
  Somewhere between rare and medium rare will be OK.
- *Is the halibut fresh?*
- *How are the scallops cooked?*
- *What is there for dessert?*

- Anything else, sir?
  No, we are fine, thank you.
- *The same for me.*
- Enjoy your meal, sir.
- How was everything, sir?
- *The meal was excellent.*
- *The sirloin was delicious.*
- *Excuse me, I have spilt something on my tie. Could you help me?*

Complaining

- *I'm afraid this dish is cold. Would you please heat it up?*
- *The meat is underdone. Would you cook it a little more, please?*
- *Excuse me. This is not what I asked for.*
- *Could you change this for me?*
- *This soup is too salty.*
- *The fish is not fresh. I want to see the manager.*
- *I asked for a sirloin.*
- *The meal wasn't very good.*
- *The meat smells off.*
- *Could you bring the complaints book?*
- *This wine is off, I think …*
- *Waiter, this fork is dirty.*

The Check (U.S.) The Bill (UK)

- *The check, please*
- *Would you bring us the check, please?*
- *Is service included?*
- *All together, please.*
- *We are paying separately.*
- *I am afraid there is a mistake, we didn't have this.*
- *This is for you.*
- *Keep the change.*

City Transportation

- *I want to go to the Metropolitan museum.*
- *Which bus/tram/subway (UK underground) line must I take for the Metropolitan?*

- *Which bus/tram/subway (UK underground) line can I take to get to the Metropolitan?*
- *Where does the number … bus stop?*
- *Does this bus go to …?*
- *How much is a one-way (UK single) ticket?*
- *Three tickets, please.*
- *Where should I get off for …?*
- *Is this seat occupied/vacant?*
- *Where can I get a taxi?*
- *How much is the fare for …?*
- *Take me to … Street.*
- *Do you know where the … is?*

Shopping

Shopping is a favorite pastime for many people all over the world. Even if you do not enjoy shopping, you may very well find that you need to buy a present for a loved one or even to survive if your luggage has been lost!

Opening Hours

- *When are you open?*
- *What time do you close?*
- *How late are you open today?*
- *Are you open on Saturday?*

Preliminary Exchanges

- Hello sir, may I help you? Hello ma'am (UK madam)
- Can I help you find something?
  No thanks, I am just looking.
- *I just can't make up my mind.*
- Can I help you with something?
- If I can help you, just let me know.
- Are you looking for something in particular?
- *I am looking for something for my wife.*
- *I am looking for something for my husband.*
- *I am looking for something for my children.*
- *It is a gift.*
- *Hi, do you sell …?*

- *I am looking for a … Can you help me?*
- *Would you tell me where the music department is?*
- *Which floor is the leather goods department on? On the ground floor (on the mezzanine, on the second floor …)*
- *Please would you show me …?*
- What kind do you want?
- *Where can I find the mirror? There is a mirror over there.*
- The changing rooms are over there.
- Only four items are allowed in the dressing room at a time.
- *Is there a public restroom here?*
- Have you decided?
- Have you made up your mind?

Buying Clothes/Shoes

- *Please, can you show me some natural silk ties?*
- *I want to buy a long-sleeved shirt.*
- *I want the pair of high-heeled shoes I have seen in the window.*
- *Would you please show me the pair in the window?*
- *What material is it?*
- *What is it made of? Cotton, leather, linen, wool, velvet, silk, nylon, acrylic fiber.*
- What size, please?
- What size do you need?
- *European sizes are different. Do you have a conversion chart or can you measure my foot?*
- *Is this my size?*
- *Do you think this is my size?*
- *Where is the fitting room?*
- Does it fit you?
- *I think it fits well although the collar is a little tight.*
- *No, it doesn't fit me.*
- *May I try a larger size?*
- *I'll try a smaller size. Would you mind bringing it to me?*
- *I'll take this one.*
- *How much is it?*
- *This is too expensive.*
- *Oh, this is a bargain!*
- *I like it.*
- *May I try this on?*
- In which color? *Navy blue, please.*
- *Do you have anything to go with this?*
- *I need a belt/a pair of socks/pair of jeans/pair of gloves …*
- *I need a size 38. (But remember sizes are different in different countries)*
- *I don't know my size. Can you measure me?*

- *Would you measure my waist, please?*
- *Do you have a shirt to match this?*
- *Do you have this in blue/in wool/in a larger size/in a smaller size?*
- *Do you have something a bit less expensive?*
- *I'd like to try this on. Where is the fitting room?*
- How would you like to pay for this? Cash/credit card
- We don't have that in your size/color.
- We are out of that item.
- *It's too tight/loose*
- *It's too expensive*
- *I don't like the color.*
- *Is it on sale?*
- *Can I have this gift wrapped?*

At the Shoe Shop

- *A pair of shoes, boots, sandals, slippers …, shoelace, sole, heel, leather, suede, rubber, shoehorn.*
- What kind of shoes do you want?
- *I want a pair of rubber-soled shoes/high-heeled shoes/leather shoes/suede slippers/boots/sandals/flip-flops/clogs.*
- *I want a pair of lace-up/slip-on shoes good for the rain/for walking.*
- What is your size, please?
- *They are a little tight/too large/too small.*
- *Would you please show me the pair in the window?*
- *Can I try a smaller/larger size, please?*
- *This one fits well.*
- *I would like some polish.*
- *I need some new laces*
- *I need a shoe-horn.*

At the Post Office

- *I need some (first class) stamps, please.*
- *First class, please.*
- *Air mail, please.*
- *I would like this to go express mail.*
- *I would like this recorded/special delivery.*
- *I need to send this second-day mail (US).*
- *Second-class for this, please.*
- *I need to send this parcel post.*
- *I need to send this by certified mail.*

- *I need to send this by registered mail.*
- *Return receipt requested, please.*
- *How much postage do I need for this?*
- *How much postage do I need to send this air mail?*
- *Do you have any envelopes?*
- *How long will it take to get there? It should arrive on Monday.*
- The forms are over there. Please fill out (UK fill in) a form and bring it back to me.

Going to the Theater (*UK* Theatre)

- Sorry, we are sold out tonight.
- Sorry, these tickets are non-refundable.
- Sorry, there are no tickets available.
- Would you like to make a reservation for another night?
- *I would like two seats for tonight's performance, please.*
- *Where are the best seats you have left?*
- *Do you have anything in the first four rows?*
- *Do you have matinees?*
- *How much are the tickets?*
- *Is it possible to exchange these for another night?*
- *Do you take checks (*UK *cheques)/credit cards?*
- *How long does the show run? About two hours.*
- *When does the show close?*
- *What time does the performance end?*
- *Is there an intermission? There is an intermission.*
- *Where are the restrooms?*
- *Where is the cloakroom?*
- *Is there anywhere we can leave our coats?*
- *Do you sell concessions?*
- *How soon does the curtain go up?*
- *Did you make a reservation?*
- *What name did you reserve the tickets under?*
- The usher will give you your program.

At the Drugstore (*UK* Chemist)

Items you might want to buy include:

- *Prescription, tablet, pill, cream, suppository, laxative, sedative, injection, bandage, Band-Aids (*UK *sticking plasters), cotton balls (*UK *cotton wool), gauze, alcohol, thermometer, sanitary napkins (*UK *sanitary towels), tampons, toothpaste, toothbrush, paper tissues, all-night pharmacy (*UK *duty chemist).*

Problems you might seek solutions for include:

- *Fever, cold, cough, headache, toothache, diarrhea, constipation, sickness, insomnia, sunburn, insect bite.*

The following phrases can be helpful:

- *I am looking for something for ...*
- *Could you give me ...?*
- *Could you give me something for ...?*
- *I need some aspirin/antiseptic/eye drops/foot powder.*
- *I need razor blades and shaving cream (*UK *foam).*
- *What are the side effects of this drug?*
- *Will this make me drowsy?*
- *Should I take this with meals?*

At the Cosmetics Counter

Items you might want to get from the cosmetic counter include:

- *Soap, shampoo, deodorant, antiperspirant, shower gel, hair spray, suntan lotion (*UK *cream), comb, hairbrush, toothpaste, toothbrush, makeup, cologne, lipstick, perfume, hair remover, scissors, face lotion, cold cream (*UK *cleansing cream), razor, shaving cream (*UK *foam).*

At the Bookshop/Newsstand (*UK* Newsagent's)

- *I would like to buy a book on the history of the city.*
- *Has this book been translated into Japanese?*
- *Have you got any Swedish newspapers/magazines/books?*
- *Where can I buy a road map?*

At the Photography Shop

- *I want a 36 exposure roll of film for this camera.*
- *I'd like some new batteries for my camera.*
- *Could you develop this film?*
- *Could you develop this film with two prints of each photograph?*
- *How much does developing cost?*
- *When will the photographs be ready?*
- *My camera is not working, would you have a look at it?*
- *Do you take passport (ID) photographs?*

- *I want an enlargement of this one and two copies of this one.*
- *Have you got a 500-megabyte data card to fit this camera?*
- *How much would a 1-gigabyte card be?*
- *How many megapixels is this one?*
- *Does it have an optical zoom?*
- *Can you print the pictures on this CD?*

At the Florist's

- *I would like to order a dozen roses.*
- *I would like a bouquet.*
- You can choose violets and orchids in several colors.
- *Which flowers are the freshest?*
- *What are these flowers called?*
- *Do you deliver?*
- *Could you please send this bouquet to the NH Abascal hotel manager at 47 Abascal St. before noon?*
- *Could you please send this card too?*

Paying

- *Where is the automatic teller?*
- *Is there a cashpoint near here?*
- *How much is that all together?*
- Will you pay in cash or by credit card?
- *Do you accept personal checks (*UK *cheque)?*
- *Next in line (*UK *queue).*
- *Could you gift-wrap it for me?*
- *Can I have a receipt, please?*

At the Barber's or Hairdresser's

At the hairdresser's, being less than fluent can result in a drastic change in your appearance, and it can take up to a few months to remedy any mistakes resulting from the failure to communicate.

If you do not trust an unknown hairdresser, "just a trim" is a polite way of avoiding a disaster.

We recommend a thorough review of your phrasebook before going to the hairdresser to familiarize yourself with key words such as: *scissors, comb, brush,* *dryer, shampooing, hairstyle, hair cut, manicure, dyeing, shave, beard, moustache, sideburns (*UK *sideboards), bangs (*UK *fringe), curl, or plait.*

Men and Women

- *How long will I have to wait?*
- Is the water OK? *It is fine/too hot/too cold.*
- *I have oily/dry hair.*
- *I have dandruff.*
- *I am losing a lot of hair.*
- *A shampoo and rinse, please.*
- How would you like it?
- Are you going for a particular look?
- *I want a (hair) cut like this.*
- *Just a trim, please.*
- However you want.
- *Is it OK?*
- *That's fine, thank you.*
- *How much is it?*
- *How much do I owe you?*
- *Do you do highlights?*
- *I would like a tint, please.*

Men

- *I want a shave.*
- *A razor cut, please.*
- *Just a trim, please.*
- *Leave the sideburns (*UK *sideboards) as they are.*
- *Trim the moustache.*
- *Trim my beard and moustache, please.*
- *Towards the back, without a part.*
- *I part my hair on the left/in the middle.*
- *Leave it long.*
- *Could you take a little more off the top/the back/the sides?*
- How much do you want me to take off?

Women

- How do I set your hair?
- What hairstyle do you want?
- *I would like my hair dyed/frosted.*

- Same color?
- *A little darker/lighter.*
- *I would like to have a perm (permanent wave).*

Cars

As always, begin with key words. You need to know *clutch, brake, blinkers (*UK *indicators), trunk (*UK *boot), tank, gearbox, windshield (*UK *windscreen) wipers, steering wheel, unleaded gas (*UK *petrol),* etc., as well as several common expressions such as:

- *How far is the nearest gas (*UK *petrol) station?* 20 miles from here.
- *In what direction?* Northeast/Los Angeles.

At the Gas/Petrol Station

- *Fill it up, please.*
- *Unleaded, please.*
- *Could you top up the battery, please?*
- *Could you check the oil, please?*
- *Could you check the tire pressures, please?*
- Do you want me to check the spare tire too? *Yes, please.*
- *Pump number 5, please.*
- *Can I have a receipt, please?*
- *I'd like 40 dollars' worth of unleaded on pump number 3, please.*

At the Garage

- *My car has broken down.*
- What do you think is wrong with it?
  *Can you fix a flat (*UK *mend a puncture)?*
- *Can you take the car in tow to downtown Boston?*
- I see …, kill the engine, please.
- Start the engine, please.
- *The car drifts to the right and overheats.*
- Have you noticed if it loses water/gas/oil?
  Yes, it's losing oil.
- Does it lose power?
  Yes, and it doesn't start properly.
- *I can't get it into reverse.*
- *The engine makes funny noises.*
- *Please, repair it as soon as possible.*

- *I wonder if you can fix it temporarily.*
- *How long will it take to repair?*
- I am afraid we have to send for spare parts.
- *The car is very heavy on gas (*UK *petrol).*
- *I think the right front tire needs changing.*
- *I guess the valve is broken.*
- *Is my car ready?*
- *Have you finished fixing the car?*
- *Did you fix the car?*
- *Do you think you can fix it today?*
- *I think my rear passenger-side tire is flat (*UK *I've got a puncture rear offside.)*
- *The spare's flat as well.*
- *I've run out of gas (*UK *petrol).*
- *It's making a funny noise.*

Parking

- *Can I park here?*
- *Where is the nearest ticket dispenser?*
- *Do you know where the nearest parking lot (*UK *car park) is?*
- *Are there any free spaces?*
- *How much is it per hour?*
- *Is the lot (*UK *car park) supervised?*
- *How long can I leave the car here?*

Renting a Car

- *I want to rent (*UK *hire) a car/van/SUV.*
- For how many days?
- Unlimited mileage?
- *What is the cost per mile?*
- *Is insurance included?*
- You need to leave a deposit.
- Can I see your driver's license (UK *driving license)?*

Asking for Directions

- *Can you tell me how to get to Milwaukee?*
- *How far is Minneapolis?*
- *Is this the road Bournemouth?*
- It is not far. About 12 miles from here.
- *Is the road good?*

- It is not bad, although a bit slow.
- *Is there a toll road between here and Berlin?*
- *How long does it take to get to Key West?*
- *I am lost. Could you tell me how I can get back to the highway?*

Going Out for a Drink (or Two)

A drink can really hit the spot after a hard day of meetings. However, ordering a drink can sometimes be complicated. Often, there is a difficult counter-question to a simple "Can I have a beer?" such as "would you prefer lager?" or "small, medium, or large, sir?" I hated counter-questions when I was a beginner. I remember how I turned beet red in a pub in London when, instead of giving me the beer I asked for, the barman responded with the entire list of beers in stock. "I have changed my mind, I'll have a Coke instead" was my response to both the "aggression" I suffered from the barman and the embarrassment resulting from my lack of fluency. "We don't serve Coke here, sir." These situations can put a damper on the most promising evening so … let's review some common expressions:

- *Two beers please, my friend will pay.*
- *Two pints of bitter and half a lager, please.*
- *Where can I find a good place to go for a drink?*
- *Where can we go for a drink at this time of the evening?*
- *Do you know any bars (*UK *pubs) with live music?*
- What can I get you?
- *I'm driving. Just an orange juice, please.*
- *A glass of wine and two beers, please.*
- *A gin and tonic.*
- *A glass of brandy. Would you please warm the glass?*
- *Scotch, please.*
- Do you want it straight (UK neat), with water, or on the rocks (with ice)?
- *Make it a double.*
- *I'll have the same again, please.*
- *Two ice cubes and a teaspoon, please.*
- *This is on me.*
- *What those ladies are having is on me.*

On the Phone

Many problems start when you lift the receiver. Even the dial tone, ring, and busy signal differ from one country to another. Phone conversations are among the most terrifying situations for non-fluent speakers because the phone has two added difficulties: its immediacy and the absence of image ("if I could see this guy I would understand what he was saying"). Do not worry, the preliminary exchanges in this conversational scenario are few. And to make matters worse, after rehearsing a conversation in your mind you may find yourself talking to an answering machine! A few years ago, most people felt ridiculous talking to a machine, but nowadays almost everyone is used to it. If you get a recorded message, do not hang up. Try to catch what the machine is saying and give it another try if you are not able to follow its instructions. There's no need to be afraid of an answering machine – most messages are much easier to understand and less mechanical than those given by "human" (and usually bored) operators.

- *Is there a public phone near here?*
- *Where is the nearest phone booth (*UK *call-box), please?*
- *This telephone is out of order.*
- *Operator, what do I dial to call China?*
- Hold on a moment … number one.
- *Would you get me this number, please?*
- Dial straight through.
- *What time does the cheap rate begin?*
- *Have you got any phone cards, please?*
- *Can I use your cell/mobile phone, please?*
- *Do you have a phone book (directory)?*
- *I'd like to make a reverse charge call to Korea.*
- *I am trying to use my phone card, but I am not getting through.*
- *Hello, this is Dr. Vida speaking.*
- *The line is busy (*UK *engaged).*
- *There's no answer.*
- *It's a bad line.*
- *I've been cut off.*
- *I would like the number for Dr. Vida on Green Street.*
- *What is the area code for Los Angeles?*
- *I can't get through to this number. Would you dial it for me?*
- *Can you put me through to Spain?*

Emergency Situations

- *I want to report a fire/a robbery/an accident.*
- *This is an emergency. We need an ambulance/the police.*
- *Get me the police and hurry.*

In the Bank

Nowadays, the widespread use of credit cards and internet banking make this section virtually unnecessary but, in my experience, when things go really wrong you may need to go to a bank. Fluency disappears in stressful situations, so if you have to solve a banking problem, you should review not only the expressions below but also the entire section in your phrasebook.

- *Where can I exchange money?*
- *I'd like to exchange 200 Euros.*
- *I want to exchange 1,000 Euros into Dollars/Pounds.*
- *Could I have it in tens, please?*
- *What's the exchange rate?*
- *What's the rate of exchange from Euros to Dollars?*
- *What are the banking hours?*
- *I want to exchange this travelers' check (*UK *cheque).*
- *Have you received a transfer from Rosario Nadal addressed to Fiona Shaw?*
- *Can I cash this check that's made out to me (*UK *bearer cheque)?*
- *I want to cash this check (*UK *cheque).*
- *Do I need my ID to cash this check (*UK *cheque)?*
- Go to the cash desk.
- Go to counter number 5.
- *May I open a current account?*
- *Where is the nearest cash machine?*
- *I am afraid you don't seem to be able to solve my problem. Can I see the manager?*
- *Who is in charge?*
- *Could you call my bank in France? There must have been a problem with a transfer addressed to myself.*

Police Matters

- *Where is the nearest police station?*
- *Can you tell me how to get to the nearest police station?*
- *I have come to report a ....*
- *I'd like to report....*
- *My wallet has been stolen.*
- *Can I call my lawyer (UK solicitor)?*
- *I have been assaulted.*
- *My laptop has disappeared from my room.*
- *I have lost my passport.*
- *I will not say anything until I have spoken to my lawyer/solicitor.*
- *I have had a car accident.*
- *Why have you arrested me? I've done nothing.*
- *Am I under caution (UK)?*
- *Am I being charged?*
- *I would like to call my embassy/consulate.*

CHAPTER 10

Chapter 10 Improving Your Pronunciation

English pronunciation is difficult for most non-native speakers. English has a complex phonological system that differs from most European languages and differs greatly from most Asian languages. However, correct pronunciation is essential – what good will knowing a lot of vocabulary and using perfect grammar do you in a presentation if nobody can understand you?

Just as it may be difficult for you to improve your overall level of English between the time you agree to do a talk and the day of the presentation, it will probably be difficult for you to make great strides in your general pronunciation skills in a short time. Nevertheless, you *must* find the time to learn to deliver your presentation in comprehensible English. It is essential to learn how to pronounce all the key words in your presentation correctly. If possible, enlist the help of a native or very proficient speaker. Ask him or her to record the text for you, and then rehearse assiduously, paying special attention to pauses and intonation.

Learning English and correct English pronunciation is a lifelong adventure. Opportunities to travel and meet people have increased dramatically. Moreover, new technologies make it possible for you to listen to and even interact with native speakers without ever leaving your own country. Your goal should not be to speak like a native speaker. Experts tell us that this objective is totally unrealistic because it is nearly impossible to attain without a sustained contact with the language before puberty. Furthermore, if you were going to attempt to speak like a native speaker, which native speaker would you emulate? English is a world language with a myriad of accents and local peculiarities. *Perhaps* it would be worth the effort to learn to speak like the inhabitants of a small town in Appalachia if you were planning to live there for the rest of your life, though the following anecdote might change your mind.

I once attended the inauguration of new medical research center in a European country. The event included back-to-back speeches by two English-speaking physicians, one from Texas in the United States and the other from Glasgow in Scotland. After each speech, the audience, most of whom did not understand a word of what the eminent doctors had said, applauded politely. Next up at the podium was the governor of the region where the research center had been built. He started out speaking in his own language "These Anglo-Saxons think that they dominate the world with their English! They think that they can go anywhere and speak their English and everyone will understand them, but nothing

J. Giba and R. Ribes, *Preparing and Delivering Scientific Presentations*,

DOI: 10.1007/978-3-642-15889-6_10,

can be farther from the truth. Only the people from Texas can understand the people from Texas, and only those from Glasgow can understand those from Glasgow. No, *their* English isn't the universal language – *my* English is the universal language because everybody can understand me wherever I go!"

And the truth is if you go to work at Harvard University in Boston, your colleagues will hail not only from all over the United States but also from all over the world. Few if any will speak like the residents of the neighborhood portrayed in Clint Eastwood's Academy Award-winning 2003 film "*Mystic River*". As we become more and more educated, we tend to lose the specific characteristics of the language we assimilated as children; the speech of most university-educated professionals has to a large extent been purged of racial, geographic, and economic class markers to become more standard and universal.

Thus, a more realistic and probably more desirable goal is to try to learn to pronounce as clearly as possible so that everyone can understand you. Make no mistake about it, this goal is challenging but achievable. Countless individuals have learned to speak English acceptably well without ever stepping foot in an English-speaking country, and as new technologies make the world smaller and smaller, it is becoming easier and easier to do so.

Obviously, it is nearly impossible to learn correct English pronunciation from a book, and the purpose of this chapter is not to teach you English pronunciation. Rather, we provide a brief description of some important aspects of English pronunciation together with some advice on how to go about improving your pronunciation.

Important Steps to Focus the Learning of Correct English Pronunciation

Become Aware of the Differences Between Your Native Tongue and English

The phonological differences between English and your mother tongue will be a major factor in your ability to learn to pronounce English correctly. Just as knowing the differences between English and your language in grammar and word-building can help you focus your learning and avoid typical mistakes in writing, knowing the differences between the phonological systems will help you overcome the specific difficulties that affect speakers of your language and avoid developing an accent that would make it very difficult for speakers of languages other than your own to understand you.

Remember that native speakers will often interpret grammar or vocabulary mistakes as a "strong accent", so be sure no eliminate or minimize these errors.

English vowels can be very difficult for speakers of most languages to master. Fortunately, correct pronunciation of vowel sounds is relatively unimportant. The pronunciation of vowel sounds varies widely among different communities of English speakers. Moreover, provided other aspects of spoken language (like grammar and vocabulary, syllable stress and intonation, and consonant pronunciation) are correct, listeners' brains will usually automatically adjust mispronunciations of vowels to the point where they may not even notice mistakes. The most important vowel sound for non-native-English speakers to master is the neutral vowel or schwa /ə/; however, although the schwa is the most common sound in English, the importance of mastering the pronunciation of this sound does not derive from the sound itself but rather from its role in intonation (see below).

English contains many "minimal pairs", that is words distinguished by a difference in a single phoneme. Some of these are vowels, but many are defined by the difference between a voiced and an unvoiced consonant produced with same position of the lips, teeth, and tongue. Searching for "minimal pairs" on the internet will yield many sites that can help you practice the ones that give you trouble.

Speakers of many languages, like Chinese, Italian, Spanish, and Vietnamese, to name just a few, also have difficulties pronouncing final consonants in English. This is a significant handicap that will require hard work to overcome. The initial and final consonants of words are crucial to comprehension.

Pronounce the Endings of Words Correctly

One area where it is definitely worthwhile to invest time and energy is the pronunciation of the final S in plurals and the third person singular of the present simple tense. This final consonant has three different pronunciations. After unvoiced consonants (those not accompanied by vibration of the vocal cords) with the exception of after /s/, /ʃ/, and /tʃ/, it is unvoiced /s/ (like the sound we make to imitate a snake), for example: *circulates* /'s3rkyə'leɪts/, *cuffs* /kʌfs/, *thinks* /θɪŋks/, *polyps* /'pɒlɪps/. After voiced sounds (those accompanied by vibration of the vocal cords) with the exception of /z/, /zh/, and /dʒ/, final S is voiced /z/ (like the sound we make to imitate a bee), for example: *allows* /ə'laʊz/, *determines* / dɪ't3rmɪnz/, *measures* /'mɛʒərz/, *films* /fɪlmz/. In both of these cases, the word ending in S has the same number of syllables as the singular or infinitive form, and the distinction between these two sounds is subtle and confusion between the two in this position is unlikely to hinder comprehension excessively. However, after the phonemes /s/, /ʃ/, /tʃ/, /z/, /zh/ y /dʒ/, the final ES is pronounced as another syllable /ɪz/, for example: *analyzes* /'ænl'aɪz ɪz/, *increases* /ɪn'kris ɪz/, *watches* /'wɒtʃ ɪz/, *proposes* /prə'poʊz ɪz/, *images* /'ɪmɪdʒ ɪz/. Confusion between this third group and the first two *will* hinder comprehension.

Another area that deserves attention is the pronunciation of the *–ed* ending of the simple past and past participle of regular verbs. This ending also has three different pronunciations. After unvoiced sounds (/f/, /k/, /p/, /s/, /ʃ/, /tʃ/ , and /θ/), *–ed* is pronounced as /t/, and the resulting word has the same number of syllables as the infinitive, for example: *assessed* /ə'sɛst/, *attached* /ə'tætʃt/, *decreased* /dɪ'krist/, *diagnosed* /'daɪəg'noʊst/, *lacked* /lækt/, *missed* /mɪst/, *peaked* /pikt/, *published* /'pʌblɪʃt/, *reduced* /rɪ'dust/. After voiced sounds (/b/, /g/, /dʒ/, /l/, /m/, /n/, /r/, /ð/, /z/ and all vowels), the *–ed* ending is pronounced /d/, and the resulting word has the same number of syllables as the infinitive, for example: *administered* /æd'mɪnəstərd/, *appeared* /ə'pɪərd/, *approved* /ə'pruvd/, *compared* /kəm'pɛərd/, *examined* /ɪg'zæmɪnd/, *fulfilled* /fʊl'fild/, *required* /rɪ'kwaɪᵊrd/. The distinction between these two sounds is also subtle and unlikely to hinder comprehension. However, after the consonant phonemes /t/or /d/, the *–ed* ending is pronounced as another, separate syllable /ɪd/, for example: *avoided* /ə'vɔɪ dɪd/, *completed* /kəm'pli tɪd/, *connected* /kə'nɛk tɪd/, *decided* /dɪ'saɪ dɪd/, *demonstrated* /'dɛmənˌstreɪ tɪd/, *detected* /dɪ'tɛk tɪd/, *extended* /ɪk'stɛndɪd/, *funded* /fʌn dɪd/, *inserted* /ɪn's3r tɪd/, *isolated* /'aɪsəˌleɪ tɪd/, *prevented* /prɪ'vɛn tɪd/, *reported* /rɪ'pɔr tɪd/, etc. Failure to distinguish between this third group and the first two is a serious error that will not go unnoticed and may hinder comprehension.

However, it is important to be aware of the exceptional pronunciation of another, separate syllable given to some participial adjectives that end in consonants other than T or D to distinguish them from their homographic verbs, for example: *aged* /'eɪdʒɪd/, *marked* /'mar kɪd/, *crooked* /'krʊkɪd/, *learned* /'l3rnɪd/, *supposed* /sə'poʊ zɪd/, *jagged* /'dʒæ gɪd/.

Learn to Deal with the Idiosyncrasies of English Orthography

One major obstacle to learning to pronounce English is the discrepancy between the spelling of English words and their pronunciation. In some other languages like Spanish, Czech, or Italian, to name just a few, you can immediately know how to pronounce any word in the language as soon as you see it; however, in English different combinations of letters can represent the same sounds and the same combination of letters can represent diverse sounds.

Be sure to take the time to learn the pronunciation of new words you come across in your reading. Listening to a recorded text while reading its transcript is one of the most efficient ways to learn correct pronunciation (more on this below). However, we do not usually have access to spoken versions of most of what we read. Maximize your learning by looking up the pronunciation of new words when you look up their meanings. Online dictionaries (see Appendix 6) will often allow you to listen to the pronunciation.

Some students find it useful to learn the International Phonetic Alphabet (IPA). In addition to enabling you to understand the pronunciations provided in writing in many dictionaries, knowing the IPA also provides you with a way to record the pronunciations of the words you learn.

While it may seem that the pronunciation is largely unpredictable from the written word, there are rules and knowing them is extremely useful. It is possible to know how to pronounce about three quarters of English words when you see them in context.

Note also that the pronunciation of Latin and Greek words is governed by a completely different set of rules than those that govern the pronunciation of ordinary English words. Appendix 1 describes the rules for pronouncing these words in English.

Homographs are words that spelled the same but pronounced differently. English has many homographs. Some of these are actually the same word pronounced differently depending on its grammatical function within a sentence. These grammatical homographs normally have a single entry in the dictionary with different pronunciations listed with the definitions for the word's different grammatical uses.

In some of these cases, the syllable stress shifts. These words are normally pronounced with the stress at the beginning when they are nouns or adjectives and with the stress at the end when they are verbs. Here is a list of the most common words in this group: *aliment, attribute, compact, compound, conduct, console, contract, contrast, implant, object, perfect, present, project, rebound, record, reject, refuse, subject, suspect, transplant.* The word *invalid* is stressed on the first syllable when it is a noun and on the second when it is an adjective.

Words that end in *–ate* are also grammatical homographs. These words are very common in medicine and science, so it is important to know how to pronounce them. The stress, normally on the antepenultimate syllable, does not change with the function of these words. Rather, it is the pronunciation of the *–ate* ending that marks the function of the word. The *–ate* ending is pronounced /eit/ (rhyming with gate, late, wait, etc.) when the word is used as a verb and with a short or neutral vowel when it used as a noun or adjective. Here is a list of some of the most common words in this group: *aggregate, alternate, approximate, associate, coagulate, duplicate, ejaculate, estimate, graduate, isolate, moderate, separate.*

There is an important exception to this rule: chemical compounds (salts of certain acids) ending in *–ate* are normally pronounced /eit/ (rhyming with gate, late, wait, etc.). Here are a few examples of words in this group: *acetate, benzoate, nitrate, phosphate, sulfate, etc.*

Another group of grammatical homographs are words in which the pronunciation of the letter S changes with grammatical function. In these cases, the S is normally pronounced like a Z (voiced S) when the word functions as a verb and with an unvoiced S when it functions as a noun or adjective. The most common words in this group are: *abuse, close, diffuse, excuse, house, reuse, use.* However, it is important to remember that in the verb construction *use to* meaning "accustomed to" or "taking place in the past but no longer in the present" is pronounced with an unvoiced S.

Last but not least, some completely unrelated words share the same spelling. These words, called heteronyms, have different meanings, origins, and pronunciations. Here are a few examples:

Lead /lɛd/ (noun and adjective)	*Lead* /lid/ (verb)
The metal Pb	To direct; the opposite of follow
Interventional radiologists wear lead aprons.	*Our group was selected to lead the project.*

(continued)

(continued)

Tear /tɪər/ (noun) Secretion of the lacrimal glands *Basal tears help lubricate the eyes and keep them free of dust.*	*Tear* /tɛər/ (verb, noun) Injure by rending, rent or fissure *Tears in the anterior cruciate ligament can take months to heal.*
Bow /boʊ/ (noun) Looped knot; device for shooting arrows *Tie it in a bow.*	*Bow* /baʊ/ (verb, noun) Bend or incline; yield *He bowed to the administration's pressure to contain costs.*
Wind /wɪnd/ (noun) Air in natural motion; gases *The patient complained of excessive wind.*	*Wind* /waɪnd/ (verb) Coil or twine about something *Take care not to wind the bandage around the arm too tightly.*
Wound /wund/ (noun) An injury involving rupture of the tegument by external forces *The surgical wound became infected.*	*Wound* /waʊnd/ (simple past and past participle verb of the verb *to wind = coil or twine about something)* *The nurse wound the bandage around the patient's head.*
Read, misread, proofread /rid/ (infinitive and present forms of the verb) *We read a lot every day.*	*Read, misread, proofread /rɛd/* (simple past and past participle of the verb) *We read a lot yesterday.*
Minute /'mɪnɪt/ (noun) 60 s *Wait a minute.*	Minute /maɪ'nut/ (adjective) Very small *There was a minute lesion in her eye.*
Number /'n∧mbər/ (noun, verb) Quantity, numeral; count *The number of complications was higher in the treatment group.*	*Number* /'n∧mər/ (comparative of the adjective *numb* = lacking sensation) *The patient said his left leg was number than his right.*
Live /laɪv/ (adjective) Living, alive; broadcast while happening *Drug concentrations were higher in live tissues.*	*Live* /lɪv/ (verb) To have life; to dwell or reside *We live in a small town near Barcelona.*
Polish /'poʊlɪʃ/ (adjective) From Poland *He was invited to address the Polish national congress.*	*Polish* /'pɒlɪʃ/ (verb, noun) To refine; to make smooth *I went to Zurich to polish my technique.*

Furthermore, there are a couple of Latin plurals that are homographs with related verbs:

Analyses /ə'næləsiz/ Plural of analysis *We do over 3,000 analyses a day.*	*Analyses* /'ænlaɪz ɪz/ (In British English) Third person singular of the verb *to analyse* in the present simple tense *This machine analyses 100 blood samples at a time.*
Diagnoses / daɪəg'noʊsis/ (plural noun) Plural of diagnosis *Three different specialists gave her three different diagnoses.*	*Diagnoses* /'daɪəg noʊzɪz/ Third person of the verb *to diagnose* in the present simple tense *Dr. House always diagnoses his patients' condition successfully.*

Concentrate on the Music of the Language

Prosody refers to the rhythm, stress, and intonation of a language. These suprasegmental (i.e., occurring at a higher level than individual sounds) aspects of language convey much more information than individual phonemes do. Like individual sounds, suprasegmental patterns vary from linguistic community to another. That's another reason why it is important to decide which variety of English you would like to learn so that you can choose an appropriate model. The information that follows refers to standard American English, not because it is somehow "better" than other varieties of English, but rather because the authors are more familiar with this variety. On the other hand, standard American English is probably the most widely spoken variety of English in the world and without a doubt the one with the greatest presence in the media. Thus, while American speakers may have difficulties understanding certain accents from other English-speaking countries, native English speakers around the world are often used to American speech from films and television programs.

Syllable Stress within Words

In English, every word of two or more syllables has at least one stressed syllable. Stressed syllables are generally longer, higher pitched, and more fully realized than unstressed syllables. Unstressed syllables are often "reduced", with the vowel losing its quality and becoming a neutral sound (schwa) or even disappearing, as is sometimes evidenced in written language (contractions).

Earlier in this chapter, we mentioned that some grammatical homophones are pronounced with the stress on the first syllable when used as nouns or adjectives and on the last syllable when used as verbs. Here are a few more guidelines to help you know which syllable to stress. Be aware that there are many exceptions, and that you can best learn word stress and other aspects of intonation "naturally" by listening.

- In acronyms that are read letter by letter, we stress the last letter. But be sure that you know how to pronounce the letters of the English alphabet!
- Most two-syllable nouns are stressed on the first syllable: *CONduct, DOCtor, SURgeon, PAtient, PRESent, LEsion*, etc.
- Most two-syllable adjectives are stressed on the first syllable: *REnal, COMmon, PREsent, CLEVer*, etc.
- Most two-syllable verbs are stressed on the last syllable: *aGREE, beGIN, conCLUDE, conDUCT, preSENT, etc.*
- Phrasal verbs are nearly always stressed on the first particle (adverb or preposition) rather than on the verb itself: check *UP* on, get *BACK*, put *THROUGH.*
- Nouns derived from phrasal verbs are stressed in the first part, whether it is the particle like in *UPtake* or *OUTbreak*, or the verb like in *MAKEup*.

- Compound nouns also receive the stress on the first word, regardless of whether they are written as a single word, a hyphenated compound, or two separate words. Some examples: *BLACKhead, NOTEbook, LAPtop, E-mail, HALF-life, TEST tube, OPERATING room, etc.*

Suffixes Can Provide the Key to Knowing Which Word to Stress

- Words ending in *–graphy* or *–scopy* are stressed on the third to the last (antepenultimate syllable): *coloNOscopy, enDOscopy, mamMOgraphy, radiOgraphy, ultrasoNOgraphy, etc.*
- Words ending in *–ic* are stressed on the next to the last (penultimate) syllable: *bioLOGic, histoLOGic, mammoGRAPHic, microSCOPic, radioLOGic, uroLOGic, etc.*
- Words ending in *–sion* and *–tion* are stressed on the next to the last syllable: *aBLAtion, aBRAsion, conCUSsion, conFUsion, etc.*
- Words ending in *–cy*, *–gy*, *-phy*, and *-ty* are stressed on the third to the last (antepenultimate) syllable: *deFIciency, biOLogy, mamMOGraphy, acTIvity, etc.*
- Words ending in *–al* are stressed on the third to the last syllable: *bioLOGical, histological, mammoGRAPHical, pathoLOGical, radioLOGical, SURGical, uroLOGical, etc.*
- Words ending in *–cious* or *–tious* are stressed on the next to the last syllable: *CONscious, inFECtious, nuTRItious, susPIcious, etc.*
- Words ending in –ate are normally stressed on the third to the last syllable, regardless of how the ending –ate is pronounced: *apPROpriate, exPECtorate, MOderate, Operate, neGOtiat, etc.*
- Words ending in *–ize* (U.K. often written *–ise*) are stressed on the third to the last syllable: *CAUterize, hyPOthesize, SYNthesize*, etc, but note: *CATHeterize, GENeralize.*
- When any of the following suffixes are added to a root word, the syllable stress of the root word does not change: *-able, -ed, -er, -est, -ful, -ing, -ment, -ness.*

Intonation

Intonation is the music of the language resulting from variations in rhythm and pitch in speech. Additionally, we use intonation to express all sorts of information beyond the meaning transmitted by words and grammar, for instance, to signal feelings and attitudes or sarcasm and irony, to name just a few examples. There are many patterns of intonation, and a complete description of all the possible patterns is beyond the scope of this book. We will look at a few of the most common patterns and try to give you some guidelines to help you learn to focus better when listening to speakers or audio material.

Before moving on to more complex patterns, it is useful to have a look at how groups of two or three words combine. In the following examples, the word in bold is stressed. In combinations of adjectives and nouns, the noun alone is stressed when the adjective is merely descriptive: A large **tumor**. Sixteen metallic **prostheses**. Male Wistar **rats.** But when the adjective (or noun in a compound) is considered to define a type, it is stressed as well. Solid **liver** lesions. **Plain**-film **radiography**. Cardiac **stress** test. **Electrophysiology** examination. Adjectives are normally stressed when they appear alone after the verb. "The **tumor** was **malignant**." **Jennifer** is **efficient"**. "The **patient** was **conscious**."

English is considered a stress-timed language. This means that the rhythm of the language comes from stressed syllables occurring at more or less regular time intervals. In order for this to happen, we have to reduce the pronunciation of unstressed syllables. This is in stark contrast to other languages like Spanish, French, Italian, Japanese, or Finnish, in which each syllable is perceived as occurring at regular intervals. When the intonation patterns from syllable-timed languages are transferred to English, the result is flattened speech that is very difficult for native speakers to understand. It is just important to use a weak or reduced pronunciation for unstressed syllables as it is to pronounce stressed syllables clearly.

In general, we tend to stress the words that carry important new information. In most cases, that means that lexical words (nouns, verbs, sometimes adjectives or adverbs) are stressed, and grammatical words (articles, pronouns, auxiliary or linking verbs, conjunctions, and determiners are not). In particular, nouns carry the weight of meaning when conveying new information, so in a sentence like "**Analgesics** relieve **pain**" the nouns are most heavily emphasized. Positive statements tend to be pronounced with a falling intonation. Notice how the stress pattern changes when we use pronouns instead of nouns. They **relieve** it. These patterns are repeated in lists "**Analgesics** include **aspirin**, **acetaminophen**, and **opiates**" "They include **aspirin**, **acetaminophen**, and **opiates"** and in more complex sentences with two independent clauses: "**Analgesics** relieve **pain**, and **sedatives** reduce **anxiety**". Introductory phrases are pronounced with a rising tone: "As we all **know**, **analgesics** can be **addictive**." As we all **know**, they can be **addictive."**

The intonation in ordinary questions rises and then drops back down at the end. Will the **analgesics** be **able** to relieve the **pain**? "Will they be **able** to **relieve** it?"

Stress timing means that the intonation is very similar regardless of whether the verb tense is very simple or very complex. The two sentences "**Surgeons** excise **tumors**" and "They **excise** them" should take roughly as long to say as "**Surgeons** have been excising **tumors**" and "They have been **excising** them" because the number of stressed syllables is the same.

Unlike in positive sentences or questions, in negative sentences the auxiliary verb is stressed. Compare "**Physicians** can prescribe **medication**" with "**Nurses can't** prescribe **medication**". This change in stress pattern is an essential element in the expression of the speaker's meaning.

We can also stress different words in sentences to shift the emphasis. The following examples show how changing the stressed word can change the meaning of the sentence.

I *don't think she should biopsy that lesion*: somebody else might think so, but I do not.

I **don't** *think she should biopsy that lesion*: somebody has understood or suggested or insinuated that I think she should biopsy that lesion, but it is not true.

I don't **think** *she should biopsy that lesion*: emphasizes that I'm not sure whether she should do it.

I don't think **she** *should biopsy that lesion*: emphasizes that she is not the right person to do the biopsy.

I don't think she **should** *biopsy that lesion*: it would be wrong to do it.

I don't think she should **biopsy** *that lesion*: emphasizes that biopsy is not the right approach – perhaps the lesion should be excised without biopsy.

I don't think she should biopsy **that** *lesion*. She should biopsy a different lesion.

I don't think she should biopsy that **lesion**. She should biopsy something else or perhaps my interlocutor did not hear me correctly and I'm repeating the word that was misunderstood.

When you stress the wrong words, you can confuse your listener. If you replace the usual pattern for a sentence like "Dr. **Darnell** learned **laparoscopy**" with "Dr. Darnell **learned** laparoscopy", your listener might understand that you are implying the second part of the complex sentence that is expressed with that stress pattern "…, but she doesn't **use** it". Likewise, if we often stress the contrast words when we should be using the basic pattern, we can sound antagonistic, as in the following examples: "Dr. James **is** a good surgeon" (as if somebody had insinuated she wasn't), "He doesn't **want** to do the operation" (as if it's purely a question of his desire, rather than for a good reason).

Make the Most of Reading and Listening

A skills-based approach to English often breaks learning down into four major areas: listening, speaking, reading, and writing. These four basic skills can be further grouped in two dimensions:

	Oral communication	*Written communication*
Comprehension (input)	Listening	Reading
Production (output)	Speaking	Writing

Listening and reading involve processing language input to extract meaning. Although you might consider these activities passive, both require active attention and information processing. A career in medicine or science requires you to read a lot in English, and most professionals have access to abundant material written in English. Moreover, most students find reading is the easiest of the four activities. The reader is free to determine the pace of language processing and to stop and think about the meaning of the text. Readers also have the opportunity to reread problematic passages, to refer back to early explanations of the concepts used or to dictionaries or other reference sources, and to analyze what is written. However, given the difficulty of English orthography, readers often learn vocabulary without having any idea of how the words are pronounced. It is important to learn not only the meaning of new words but also their pronunciation. As mentioned above, new technologies make it easy to hear native speakers pronounce words.

Reading aloud is a very good way to improve your pronunciation. First, it helps you realize which words you do not know to pronounce. You should pay attention to problematic aspects of pronunciation discussed elsewhere in this chapter, like pronouncing the beginnings and ends of words clearly, pausing in the appropriate places, getting the word stress right, and using the correct sentence intonation by focusing on saying the important syllables clearly while reducing the length, loudness, and pitch of unstressed syllables.

One of the most powerful strategies to improve your pronunciation is combining listening and reading in a single activity. This will help to strengthen your knowledge of the way English sounds are expressed in writing. Remember that even though there is no one-to-one correspondence between sounds and the letters that represent them in written English, the pronunciation of over three-quarters of the words in English follows regular rules.

It used to be relatively difficult to find suitable material available in both written and spoken form, but nowadays, there are many sources of dual-input texts. Apart from traditional English language teaching textbooks with listening and tapescripts, there are many internet sites devoted to English language teaching that provide you with the opportunity for listening practice. A wide variety of subject matters are covered for learners at different levels. Many sites provide the scripts and accompanying exercises.

An ever-increasing amount of audio and audiovisual material providing medical and scientific information for both professionals and laymen is also available on the internet. Moreover, including the terms podcast, vodcast, webcast, webinars, or video together with the subject matter of your presentation can turn up interesting links to presentations recorded at congresses, continuing medical education material, training videos, and much more. Many scientific journals also include some spoken content, and some of these also make the transcripts of this content available. Appendix 6 lists just a small sample of useful websites for improving your listening comprehension and pronunciation of medical English.

In many countries, you can watch English language programs and films on television with subtitles in your own language or in English. Whenever possible, you should choose to read the subtitles in English. Television audiences' fascination with life-and-death situations ensures that hospitals figure prominently in many films and programs. American series based in hospitals, like *Grey's Anatomy, Scrubs, and House* or British series like *Holby City* can also help you learn some colloquial hospital vocabulary. Whether or not these shows are available on television in your country, you can often purchase DVDs of an entire season or watch many episodes for free on the internet.

Audiobooks are another great source of language input. Be sure to obtain a copy of the printed work as well. Nonfiction is generally easier and more useful than fiction, because actors often dramatize the content of works of fiction, resulting in unnatural speech patterns. No matter what you read and listen to, be sure that the recording is human speech – computerized voices like the one produced on Amazon's Kindle are not useful at all for our purposes. Many libraries give members free online access to this type of material.

Cultivate an Attitude That Is Conducive to Learning

Your attitude can greatly influence your success in learning correct English pronunciation. Here are a few general tips to help you learn to pronounce better in English:

1. Decide that English pronunciation is important for you. Be determined to improve your pronunciation. Commit yourself to working on problematic aspects.
2. Try to imitate native speakers. Do not be afraid of sounding different or strange: English sounds *are* different and funny. It may be useful to practice exaggerating the English or American accent in your own language to help you become aware of the differential traits between the two languages and start to lose any embarrassment you might have about speaking in English. Shyness or embarrassment about the way you sound is the greatest obstacle you must overcome to make any progress whatsoever in speaking a language. Sometimes non-native-speaking physicians know how to pronounce a word correctly but are a bit ashamed of doing so, particularly in the presence of colleagues from their own country. Do not be ashamed of pronouncing correctly, regardless of the nationality of your interlocutor.
3. Pay attention to what is happening in your throat and mouth when you speak English. Producing the sounds of English involves using your vocal apparatus in ways you might not be accustomed to. You may experience some discomfort: your muscles may even become stiff or sore – this is a sign that you are working hard – persevere and you will get used to using them. With time, any discomfort or feeling of strangeness will disappear, and producing English sounds will become automatic. Some experts even postulate a "muscle memory" for sounds.
4. Do not worry about having a thick accent. Accents do not matter much, as long as people can understand you. The idea is to communicate what you think or feel, rather than to impress a linguist or speech therapist.
5. Rehearse common expressions and standard collocations in both everyday conversational and biomedical scenarios. Practicing simple phrases like "Do you know what I mean?" or "Would you do me a favor?" and "Who's on call today?" "Can you give me a hand with this?" will help build your confidence and make you more fluent.

Practical Steps to Improve Pronunciation

1. *Slow down.* No matter how good your pronunciation is, it will be more difficult for people to understand you if you speak quickly. Speaking slowly gives your lips and tongue time to reach the correct positions to produce the sounds of English. Speaking slowly will also help you relax and breathe, both of which are important for communicating well.
2. *Speak up.* Whether you are talking to a full conference hall or a single individual, make sure that nobody needs to struggle to hear you. If people often ask you to repeat or clarify what you have said, the problem may lie in the volume of your voice rather than in your inability to speak well. If people in the audience tend to lean forward, you may not be projecting voice enough. Speaking too softly might make you seem unsure of yourself and detract from your message; speaking at the appropriate volume will make you sound confident and in control, and these qualities are very important in many fields.
3. *Practice.* Practice makes perfect. Never pass up an opportunity to speak English. Try to listen to and speak English every day. Choose a short passage from an interesting recording and listen to it repeatedly. Try to remember useful sentences from the recording and practice saying them aloud to yourself. If your English is already quite good, consider joining Toastmasters International www.toastmasters.org to improve your public speaking skills.
4. *Get feedback.* Record yourself often. Try recording short texts from the transcripts of material recorded by a native speaker and comparing the two recordings. This will help you notice mistakes in all aspects of pronunciation. Pay special attention to syllable stress and intonation, pausing, and liaisons between words. If you can have a native speaker record the text of your presentation, this technique can work wonders in a relatively short time. A dedicated tape recorder is probably unnecessary, as many digital devices, including most computers, mobile phones, and cameras, have recording capabilities. As mentioned in the chapter on delivery, video recording of your presentation can also help with other aspects like posture, gestures, and facial expressions.

A Few Final Words

Preparing an oral communication for an international congress is an exciting challenge. A successful presentation will give you a wonderful sense of accomplishment. But remember, this is only the first step.

Most of the presentations at international congresses can form part of the raw material for a journal article or book chapter. Writing up a formal export is also a lot of work, but much of the groundwork will have been laid laid during the preparations for your presentation.

You might also find yourself eagerly looking ahead to your next project and submitting new abstracts for talks in English. If you were truly outstanding, someone might be contacting you to invite you to speak at another event.

Take the opportunity to follow up on some of the professional contacts you made during the congress. Write a short email to thank the chairperson of the session, and don't forget to mention any special help he or she gave you. If you found some of the other talks in your session interesting, let the speaker know, in person at the congress or by email afterward or, preferably, both.

Never forget how much work you put into your success, and if the opportunity arises somewhere down the line, extend a helping hand to someone who is just starting out.

CHAPTER 11

Chapter 11 Appendices

Appendix 1: Latin and Greek in English

About 60% of the words in English have Latin or Greek roots. In science and technology, estimates are as high as 90%. Although we refer to Latin and Greek as "dead languages", in a sense they are very much alive today. There is an international scientific vocabulary in which different elements from classical languages are combined to form new words and new very specific meanings are assigned to words that existed with general meanings in their original forms and usage. Every modern language has its version of the international scientific vocabulary – the spellings and pronunciations are adapted to fit the characteristics of the language and make it easier for native speakers to deal with.

English has a long tradition of using Greek and Latin words for science. This has led to a sort of sublanguage, a language within a language, although like other aspects of language use, our use of Greek and Latin is slowly evolving. In general, English is more conservative in the use of these forms, retaining original plural forms where other languages apply their own typical rules for plural formation. This appendix is divided into two parts: the first presents some rules to help you appreciate how we form plurals of this kind of word in English, and the second provides some guidelines about their pronunciation.

Always check both the plural form and the pronunciation of these words in a reliable source before including them in your presentation.

Part 1: Forming Plurals

The following guidelines can help you get used to guessing the plural forms.

1. Most words ending in *–us* change to *–i*:

alveolus	alveoli
bronchus	bronchi
nevus	nevi
sulcus	sulci
Exceptions to *–us* changing to *–i*:	
corpus	corpora

(continued)

J. Giba and R. Ribes, *Preparing and Delivering Scientific Presentations*,
DOI: 10.1007/978-3-642-15889-6_11, © Springer-Verlag Berlin Heidelberg 2011

(continued)

ductus	ductus
genus	genera
plexus	plexuses

2. Words ending in *–um* change to *– a*:

acetabulum	acetabula
diverticulum	diverticula
hilum	hila
septum	septa

3. Most words ending in *–a* change to *– ae*:

areola	areolae
fenestra	fenestrae
lamina	laminae
sequela	sequelae

4. Words ending in *–ma* change to *–mata* or *–mas*:

carcinoma	carcinomas (carcinomata)
chiasma	chiasmata
sarcoma	sarcomas
stoma	stomata

5. Some words ending in *–is* change to *– es*:

anastomosis	anastomoses
crisis	crises
metaphysis	metaphyses
metastasis	metastases

6. Some words ending in *–is* change to *– ides*:

arthritis	arthritides
dermatitis	dermatides
glottis	glottides
iris	irides

7. Words ending in *–x* change to *– ces*:

apex	apices
calyx	calyces
cervix	cervices
thorax	thoraces

8. Words ending in *–cyx* change to – *cyges*:

coccyx	coccyges
larynx	larynges
pharynx	pharynges
salpinx	salpinges

9. Words ending in *–ion* change to – *ia*:

criterion	criteria
ganglion	ganglia
mitochondrion	mitochondria
phenomenon	phenomena

Part 2: Pronunciation

We will look briefly at two aspects of the pronunciation of Latin and Greek words in English: the sounds represented by different letters and syllable stress. We will see that these two aspects are interrelated: syllable stress determines the pronunciation of certain vowels and the presence of certain combinations of letters determines syllable stress.

Some of these rules are complicated and it is not very practical to memorize them. Learning rules and applying them is not the best way to learn pronunciation; however, familiarity with these rules can help improve your intuitive understanding of pronunciation and ability to guess the pronunciation of unfamiliar words. As in other areas of English, exceptions abound and you should always check the pronunciation of the words you include in your presentation in a reliable source.

The sounds of letters:

The digraphs AE and OE are treated as the letter E. Especially in American English, AE and OE are often now written simply as E. Thus, a C before AE or OE is pronounced as though followed by E, i.e., as an S: Caesar, caecum, coelom, etc. Similarly, a G before AE or OE is pronounced as though followed by E, i.e., like a J: algae, rugae, etc.

The letter E can have two pronunciations:

Long E rhymes with the English words me, we, be, tree, etc.

Examples: ameba (amoeba), anemia (anaemia), cecum (caecum), ether (aether), etiology (aetiology), fetus (foetus), etc.

The AE ending for plurals and group names is generally pronounced as a long E: genus of yeasts causing thrush, Candidae, larvae, etc. A notable exception in medicine is vertebrae, in which most people tend to pronounce the last syllable with a long A.

Short E rhymes with the English words met, wet, set, etc.

Examples: aesthetic (esthetic), hemorrhage (haemorrhage), esophagus (oesophagus), estrogen (oestrogen), etc.

The combination *CH* is pronounced as K:

chondrosarcoma, cholera, cholelithiasis, Chlamydia, chromatolysis

TH is always pronounced unvoiced, as in

thalamus, thorax, thymus, thyroid, thalassemia, thalidomide

A final vowel is never silent.

fungi = "FUNJ-eye," i as in alibi
cocci = "COCKS-eye," i as in alibi
ovale = "oh-VAY-lee," e as in Daphne,
stapes = "STAY-peez" e as in Achilles, meninges, diabetes

Some initial consonants are silent when followed by a consonant.

pterygoid = "TER-i-goid"
pseudocyst = "SOO-do-cist"
pneumonia = "New-MON-ia"
gnathous = "NATH-ous"
Phthirus = "THI-rous"

Words of two syllables are stressed on the first syllable:

Femur = "FE-mur"
Sinus = "SI-nus"
Dermoid = "DER-moid"

The vowel of the first syllable is short if followed by two or more consonants:

Fossa (o as in cot)
Fascia (a as in cat)
Septum (e as in met)
Pinna (i as in mit)

The vowel of the first syllable is long if followed by a single consonant:

Coma
Ramus (a as in gate)
Femur (e as in me)
Iris (i as in hi)

Words of more than 2 syllables are stressed on the penultimate syllable if it is followed by two or more consonants.

Maxilla = "mac-ZILL-a"
Patella = "pa-TELL-a"
Medulla = "med-ULL-a"
Cerebellum = "ser-e-BELL-um"

Words of more than 2 syllables are stressed on the penultimate syllable if its vowel is long.

Foramen = "fo-RAY-men"
Ureter = "you-REE-ter"

Duodenum = "du-oh-DEE-num"
Pectoralis = "pec-to-RAY-lis"

Words of more than 2 syllables are stressed on the antepenultimate syllable in other cases.

Esophagus = "es-OFF-ag-us"
Parenchyma = "pa-REN-kim-ma"
Epiphysis = "e-PIF-is-is"
Stomata = "STOM-ah-tah"

The vowel is long if it is U.

Humerus = "HUM-er-us"
Numeral = "NUM-er-al"
Clenbuteral = "clen-BUT-er-al"

The vowel is long if it precedes a vowel.

Coitus = "CO-i-tus"
Iodine = "EYE-o-din"
Methionine = "ma-THI-o-neen"

The stressed vowel is long if it is A, E or O followed by a single consonant, then two or more vowels, of which the first is E, I or Y.

Media = "MEE-dee-ah"
Splenius = "EYE-o-din"
Radius = "RAY-dee-ahs"
Phobia = "FO-bee-ah"

Note the "SH" sound that may be given to c and t followed by i:

dementia = "de-MEN-sha"
motion = "Mo-shun"
acacia = "a-KAY-sha"
species = "SPEE-shees"

Some double consonants (and "mute" consonants followed by l or r) are treated as single consonants (e.g., TH, PH, CH, BR, DR, TR, PL, QU):

Sacrum = "SAY-crum"
Mitral = "MY-tral"
Nigra = "NY-gra"
Vertebra = "VER-te-bra"
Agnatha = "AG-na-tha"

The letter X is treated as two separate consonants because it has a "KS" sound:

Axis (short "a" as in cat)
Toxin, antioxidant (short "O" as in cot)
Elixir (short "i" as in kit"

Certain contractions ending in -ic, -id, and -it retain the vowel quality of the original:

Gravid, (a as in cat) from gravidus
Tropic, (o as in cot) from tropicus Cephalic, (a as in cat) from cephalicus
Hepatic, (a as in cat) from hepaticus

Stress of compounds varies with suffix. -OSCOPY, stressed third to last syllable:

Endoscopy = en-DOS-co-pee
Colonoscopy = co-lo-NOS-co-pee
Microscopy = my-CROS-co-pee
Arthroscopy = ar-THROS-co-pee

Stress of compounds varies with suffix. -IC, stressed second to last syllable:

Endoscopic = en-do-SCO-pic
Colonoscopic = co-lo-no-SCO-pic
Microscopy = my-cro-SCO-pic
Cephalic = Se-FAL-ic

Stress of compounds varies with suffix. -ulum, stressed third to last syllable:

Diverticulum = dee-ver-TIC-u-luhm
Speculum = SPEC-u-luhm
Pendulum = PEN-du-luhm

Appendix 2: British and American Spelling

Most differences in spelling between British and American English are systematic and easy to predict. Be sure to select the appropriate variety of English in PowerPoint when preparing your slides and check any words you are not sure of in a reliable reference source.

United States	United Kingdom
O→	**OU**
Colour	Color
Tumor	Tumour
Behaviour	Behavior
ER →	**RE**
Center	Centre
Meter	Metre
Caliber	Calibre
E→	**AE**
Anemia	Anaemia
Anesthesia	Anaesthesia

(continued)

(continued)

Cecum	Caecum
Etiology	Aetiology
Hematoma	Haematoma
Pediatric	Paediatric
E→	**OE**
Celiac	Coeliac
Diarrhea	Diarrhoea
Edema	Oedema
Esophagus	Oesophagus
Estrogen	Oestrogen
Fetus	Foetus
Z→	**S** *However, note that the use of Z is increasingly common in British English*
Analyze	Analyse
Catheterize	Catheterise
Criticize	Criticise
Organization	Organisation
Visualization	Visualisation
L→	**LL**
Beveled	Bevelled
Traveler	Traveller
Labeling	Labelling
LL→	**L**
Enrollment	Enrolment
Fulfill	Fulfil
Skillful	Skilful
Miscellaneous	
Analog	Analogue
License (verb and noun)	Licence (*noun – the verb is spelled license*)
Program (*all types*)	Programme (*for congresses, concerts, etc. But computer program*)
Practice (*verb and noun*)	Practise (*verb – the noun is spelled practice*)
F→	**PH**
Sulfur	Sulphur
Sulfonamide	Sulphonamide
Medical compounds tend to be written without hyphens	**Medical compounds tend to be written with hyphens**
Hepaticoduodenostomy	Hepatico-duodenostomy
Sternopericardial	Sterno-pericardial

Appendix 3: Verb Patterns

Verbs that can only be followed by the infinitive + to:
agree, aim, appear, arrange, ask, beg, claim, condescend, consent, dare, decide, demand, deserve, desire, endeavor, expect, fail, happen, hesitate, hope, intend, learn, manage, offer, plan, pretend, proceed, promise, seem, strive, swear, tend, threaten, and want.

Verbs that can only be followed by the gerund:
admit, anticipate, appreciate, avoid, complete, consider, delay, defend, deny, discuss, enjoy, finish, imagine, involve, mention, mind, miss, postpone, practice, put off, quit, recall, recollect, recommend, report, resent, resist, resume, risk, suggest, tolerate, and understand.

Verbs that can be followed by the infinitive or gerund with little or no difference in meaning:
afford, bear, begin, cease, commence, continue, dread, hate, like, loathe, neglect, prefer, propose, start, and undertake.

Verbs that have different meanings depending on whether they are followed by the infinitive or gerund:

	Verb + infinitive	Verb + gerund
attempt, try	Make an effort to do something	Do something to see if it works
	Please try to be punctual.	*I'm sorry. I've tried everything. I've tried setting two alarm clocks, skipping breakfast, coming by train, but nothing works.*
	The interventional radiologist tried to occlude the bleeding vessel.	*He tried using coils; when that didn't work, he tried adding gelfoam pledgets.*
remember	Keep something in your conscious memory so you can do it	Do something, then bring it back to your conscious memory
	Please remember to phone the referring physician when you have sent the report.	*I remember reading an article about this in the NEJM, but I don't remember when.*
forget	Not do something because you didn't remember you had to do it	Do something, then not remember having done it
	The patient forgot to fast, so the examination had to be postponed.	*The patient was sedated so he forgets undergoing the procedure.*
go on	To accomplish at a later time	To continue
	She went on to become the chief of the department.	*She went on making mistakes until she was expelled from the program.*
mean	To have the intention to do something	Involves doing something
	She meant to tell you herself but she was called out on an emergency.	*Being a physician means working nights and weekends.*

(continued)

(continued)

regret	Feel bad about something you have to do before you do it *We regret to inform you that your article has not been accepted for publication.*	Feel bad about something after actually doing it (or failing to do it) *She regrets not applying for the job.*
stop	End (an unspecified action) to begin another action *The technologist stopped (image acquisition) to administer contrast.*	End an action that is in progress *My cardiologist only stopped performing fluoroscopy last year!*

Verbs that are followed by either an indirect object + a full infinitive or by a noun (or the gerund, which is the noun form of the verb):
Advise, allow, encourage, forbid, permit, recommend

Verbs that can be followed by an indirect object + a full infinitive but not by a noun or gerund:
Enable, force, get (=arrange for, persuade), invite, order, persuade, remind, teach, tell, warn

Verbs followed by an object and the infinitive without to:
Help*, let, make, see, hear
Note: Let is not used in the passive (use *be allowed to* instead). The other verbs require *to* to be used in the passive: *She was made to repeat the procedure until she could do it perfectly*. This usage is very formal and unusual.

*Help can also be used with the full infinitive in the active voice.

Appendix 4: Numbers

Verbalizing and Pronouncing Numbers and Related Terms

It is said that numbers are the only universal language, but numbers in spoken language are anything but universal. Numbers are crucial to science, and few scientific presentations would be possible without the presenter having to say at least a few numbers out loud. The following tips are intended to help you deal with this important issue.

Some Vocabulary Related to Numbers

The word "billion" in scientific English always refers to 10^9 (never to 10^{12}, which is referred to as a trillion).

The terms *hundred, thousand, million*, etc. are always used in the singular when we refer to a specific number. For example, we would say *"we studied three hundred patients with acute lung injury*", "*Our hospital serves a population of two hundred thousand inhabitants*", and "*It is estimated that between fifty and one hundred million people died in the 1918 flu pandemic.*" However, when we use these numbers as units of counting, the plural forms can be used. For example, "*Millions of people have already died of AIDS*", "*Thousands of people are killed or maimed in automobile accidents every year*", or "*Hundreds of new drugs come on the market every year*".

A "couple" refers to two items of the same class: *We had a couple of hours to visit the town before we went to the airport. A couple of Korean doctors asked me to send them a copy of my presentation.* A couple is also used to refer to two people involved in a romantic relationship: *The elderly couple asked for information about living wills.*

A "pair" refers to two identical, similar, or corresponding items that are matched for use together: *a pair of earrings/gloves/shoes, a pair of identical twins*. A pair can also refer to something seen as being composed of two parts: *a pair of scissors/forceps/trousers.*

The following terms are often used to refer to nonspecific quantities: "a few", "several", "a number of", "many", and "a lot". The actual numbers behind these expressions can vary widely depending on the context. It is also preferable to be specific in scientific reporting.

Apart from ranking and ordering (*the first third of the descending colon*), we use ordinal numbers for naming:

The days of the month: *the fourth of July, the thirty-first of January, etc.*

Floors of buildings: *the lab is on the second floor*

Streets and avenues: *the hospital is located at the corner of Fifth Street and Twelfth Avenue*

The denominator in a fraction: *two-thirds, three-fifths, etc.*

Expressing Numbers in Writing

When preparing slides, remember that in English, we use periods not commas for decimal points. Thus, one million is written 1,000,000 (never 1.000.000) and five one-hundredths is expressed as decimal like this 0.05 (never 0,05).

As mentioned in the section on punctuation above, when we write the numbers 21 to 99 in words, we put a hyphen between the tens and ones columns (two hundred and forty-eight, sixty-four), and we also put a hyphen between the numerator and denominator when we write out fractions as words (2/3 = two-thirds).

We normally write out the word to express the number in the following cases:

At the beginning of a sentence: *Twenty-one patients died.*

For single-digit numbers: *Complications occurred in six cases.*

When two numbers are juxtaposed: *Rats underwent three 5-second shocks.*

We always express numbers with numerals in the following cases:

When written together with an abbreviated unit of measurement: *10 mL, 3 kg*
In series in which one or more of the elements is greater than ten: *boys aged 3, 7, and 12 years old; 5, 10, and 15 min after administration*.
When written within parentheses or in tables or references: *(5–9)*.

It is a good idea to repeat the % symbol and units of measurement in ranges or series: *15% to 28%; (3 kg–10 kg)*.

We put a hyphen between a numeral and "fold" but not between the number written as a word and "fold": *10-fold* but *tenfold*.

Ordinal numbers written in numerals use the abbreviations st for first *31st = thirty-first*, nd for second *42nd = forty-second*, rd for third *53rd = fifty-third*, and th for the rest *12th = twelfth, 25th=twenty-fifth*.

Expressing Numbers and Mathematical Symbols in Oral Communication

Be sure to distinguish clearly between the pronunciation of the "teens" and multiples of ten. Except in counting, thirteen, fourteen, fifteen, sixteen, seventeen, eighteen, and nineteen are stressed on the last syllable. Contrast this with thirty, forty, fifty, sixty, seventy, eighty, and ninety, which are stressed on the first syllable. Obviously, it is also very important to pronounce the final N. Remember, too, that in contrast to the I in five, the I in fifteen and fifty is short.

It is important to know when we use cardinal numbers (one, two, three, …) and when we use ordinal numbers (first, second, third,…). For example, unlike in many other languages, dates are spoken with ordinal numbers in English. Thus, *October 12* is said *"October twelfth"* or *"the twelfth of October"*. Years are cardinal numbers, but pronounced in groups of two; for example, 1960 = *nineteen sixty*, 1776 = *seventeen seventy-six*. However, the presence of zero changes this rule slightly: 1803 = *eighteen oh three*. This is actually a shortened form of the more formal and old-fashioned "*nineteen hundred sixty*", "*seventeen hundred and seventy-six*", and "*eighteen hundred and three*". This more formal form is still used, especially with the first decade of a century. Following this logic, we have started to refer to the first years of the new millennium by saying "*two thousand and …*". We are expected to revert back to grouping in twos with the year 2010 "*twenty ten*".

We use cardinal numbers to refer to decades, as in "*the sixties*" or "*the nineteen sixties*", but we use ordinal numbers to refer to centuries "*the twentieth century*" = 1900–2000.

We use cardinal numbers with percentages, but ordinal numbers with percentiles. Thus, we would say "*seventy percent of the rats developed edema*", but "*the diameter of the infant's head was in the ninety-eighth percentile*".

Decimals are read using the word "point" and listing the numbers after the decimal point one by one. Here are a few examples: 3.141519 = *three point one four one five one nine*; 342.01 = *three hundred and forty-two point oh one*; 1.065 = "*one point oh six five*"; and 0.005 = "*point oh oh five*". *His temperature was 38.6°C = thirty-eight point 6 degrees Celsius.*

Fractions are spoken by using a cardinal number for the numerator and an ordinal number for the denominator; the exceptions are "*half*" (or "*halves*") when the denominator is two, and sometimes "*quarter(s)*" when the denominator is four. The denominator is singular or plural depending on whether the numerator is one or another number. Here are a few examples, 1/3 = *one-third*, ¼ = *one-fourth* or *a quarter*, 2/3 = *two-thirds*, and 5/16 = *five-sixteenths*. Larger fractions in scientific presentations are likely to be expressed as proportions or percentages.

Proportions, often written as fractions, are verbalized with the prepositions "in", "of", or "*out of*". For example, "one in three (1/3) people will contract cancer". "*Sixty-five of sixty-eight patients underwent MR angiography.*"

Ratios, for example 3:1, are verbalized as "*three to one*". "*The ratio of men to women affected is two to one*".

Common mathematical operators are verbalized as is in the following examples:

Sums: 2 + 2 = 4 "*two plus two equals four*"; we also say "*the sum of two and two is four*"

Difference: 10–5 = 5 "*ten minus five equals five*" or "*the difference between ten and five is five*" or "*five from ten is five*".

Product 4 x 3 = 12 "*four times three equals twelve*" or "*four multiplied by three equals twelve*"

Quotient 24 ÷ 3 = 8 "*twenty-four divided by three equals eight*" or "*three divides into twenty-four eight times*" "*the quotient of twenty-four and three is eight*"

Here is the way we express some common mathematical symbols orally:

± *plus or minus*	10^9 *ten raised to the ninth power*; often shortened to *ten to the ninth*
< *less than*	≤ *less than or equal to*
> *greater than*	≥ *greater than or equal to*
□ *the square root* of	Δ *the change in*

Remember that Greek letters are pronounced as English words; thus, Π is pronounced "*pie*", which rhymes with "*my*".

Appendix 5: Words Often Confused – Similar Spellings

Abominable	Abdominal
Affect (verb: have an effect upon)	Effect (noun: consequence)
Affect (noun: conscious feeling or emotion)	Effect (verb: to bring about)
Alternately (following by turns)	Alternatively (choice between two possibilities)

(continued)

(continued)

Altogether (completely)	All together (not separately)
Asses (donkeys or buttocks)	Assess (evaluate)
Baring (exposing)	Barring (excluding)
Beside (next to)	Besides (In addition to)
Bind (join together)	Blind (unable to see)
Break down (verb)	Breakdown (noun)
Breath (noun)	Breathe (verb)
Case	Cause
Check	Cheek
Continual (intermittent)	Continuous (uninterrupted)
County	Country
Creatine	Creatinine (end product of creatine metabolism)
Farther	Father
Farther (limited to distance)	Further (can be figurative)
Filled (made full)	Filed (archived)
Forgo (do without)	Forego (go before)
Heat	Heart
Humerus (bone)	Humorous (funny)
Ileum (part of small intestine)	Ilium (part of hip bone)
Insect	Insert
Its (possive adjective)	It's (contraction of "it is")
Lactase (enzyme)	Lactose (sugar)
Larger (bigger)	Lager (beer)
Lose (opposite of win, gain, or find)	Loose (opposite of tight)
Mucus (noun)	Mucous (adjective)
Muscle	Mussel (bivalve mollusk)
On	One
Patent	Patient
Predominate (verb)	Predominant (adjective)
	Predominately (adverb)
Principal (main)	Principle (rule, generalization)
Prostate (gland)	Prostrate (prone)
Quit	Quite
Quiet	
Scarring (cicatrization)	Scaring (frightening)
Secrete	Secret
Severe (adjective: very bad in degree or extent)	Sever (verb: cut off)
Spit (expel from the mouth)	Spite
Stationary (unmoving)	Stationery (paper for writing)
Strike	Stroke

(continued)

(continued)

Suit (clothing or to fit)	Suite (room)
There	Their
Though	Through
Trough	
Very	Vary
Ward (section of the hospital)	Wart (cutaneous elevation caused by papillomavirus or benign conditions with similar appearance)
Weak (opposite of strong)	Week (7 days)
Weighed (measured the weight of)	Weighted (adjusted to reflect value or proportion)
Where	Were

Appendix 6: Useful Websites

From dictionaries and other reference sources to audiovisual files and interactive consultation, the internet contains a wealth of material that anyone who needs to prepare a presentation in English would be foolish to ignore.

Apart from searching for standard references through sources like MEDLINE, you should also consider researching related background material to help you become familiar with the language involved in your topic. Examining audiovisual material directed at patients is a good way to improve your English while learning expressions and pronunciations directly or indirectly related to the topic of your talk.

There can be no excuse for not knowing how to pronounce any of the words you use in your presentation, as many online dictionaries allow you to listen to the pronunciation of the words they list and other sites allow you to submit words to be pronounced by native speakers.

Especially interesting are opportunities to listen and read simultaneously or to watch and listen to (and often read as well) speeches and scientific presentations. This is a great way to work on the crucial aspects of intonation and presentation skills.

Moreover, new technologies allow you to download material from the internet and use it whenever and wherever you like on a portable computer, iPOD or similar MP3 or MP4 player, iPAD or other portable device. The time you spend on learning about these possibilities is a truly worthwhile investment.

Space allows us only to present the tip of this fascinating iceberg, but you will surely find more relevant material by using the keywords video, webcast, vodcast, vidcast, podcast, or presentation together with some of the key terms from your topic.

Good Places to Start Searching for Relevant Audiovisual Material

Youtube http://www.youtube.com features videos and channels about everything under the sun. Material is available to watch online only.

iTunes http://www.apple.com/itunes/store has movie rentals, movies, TV shows, podcasts, audiobooks, iPod games, and songs to buy as well as free downloadable material. *iTunes U* gives you free access to some of the world's best thinking – lectures from MIT, labs from Stanford, films from the MOMA, and more.

Google video search http://video.google.com allows you to use key terms to search for videos hosted at Google and videos hosted anywhere on the WWW.

The Health Library Wiki http://hlwiki.slais.ubc.ca/index.php/Podcasts_and_Videocasts has a wealth of information about podcasts and vodcasts for healthcare professionals and patients.

The Medical College of Georgia http://www.lib.mcg.edu/resources/podcasts.php has an interesting list of online audiovisual resources for medical professionals.

Dictionaries That Allow You to Listen to the Pronunciation of Words

www.howjsay.com Online pronunciation dictionary. Also available for iPhone and BlackBerry.

www.forvo.com This website allows you to listen to the pronunciation of words. Native speakers' sex and geographic location are displayed. Best of all, you can submit words to be pronounced if you cannot find what you're looking for in their databases.

The Merck Manuals Online Medical Library

http://www.merck.com/mmhe/index.html contains a wealth of information. It also has a pronunciation dictionary http://www.merck.com/mmhe/resources/pronunciations/index/a.html that allows you to listen to the pronunciation of thousands of medical words.

Wisc-Online is a digital library of Web-based learning resources developed primarily by faculty from the Wisconsin Technical College System.

https://www.wisc-online.com/Objects/ViewObject.aspx?ID=GEN504 provides a "Medical Terminology Pronunciation Juke Box" organized into the following categories: blood & other body fluids, body structure, cardiovascular & lymphatic systems, digestive system, endocrine system, integumentary system, medicine & specialties, muscular & skeletal system, nervous system, reproductive system, respiratory system, urinary system).

http://education.yahoo.com/reference This site contains searchable reference materials, including *The American Heritage Dictionary* (with spoken pronunciation), *Roget's Thesaurus*, *The American Heritage Spanish-English dictionary*, and *Gray's Anatomy of the Human Body.*

English as a Foreign Language (EFL) Teaching Sites

http://www.fonetiks.org is focused on the pronunciation of sounds. It provides online pronunciation guides to 9 varieties of the English language, with sound and pronunciation samples by over 40 native speakers. It offers a variety of English dialogues, an interactive reading course and many other activities which are available on the web page.

Randall's ESL Cyber Listening Lab http://www.esl-lab.com/ provides general listening quizzes; everyday conversations (adult and children's voices) graded as easy, medium, or difficult; listening quizzes for academic purposes; lectures, interviews, and conversations graded as medium, difficult, or very difficult; language and life-skills tips with audio and discussion questions; and long conversations with RealVideo, classified as easy, medium, or difficult.

Medical English Multimedia Course http://www.englishmed.com contains animated dialogues for doctors, nurses, pharmacists, and general medical staff, a resource center with thousands of exercises, and a forum for interaction with other learners and teachers.

BBC Learning English http://www.bbc.co.uk/worldservice/learningenglish/index.shtml has a wealth of interesting material to help students.

The *English Language Listening Lab Online* http://elllo.org provides over 1,000 listening activities designed especially for EFL students. Most activities include images, an interactive quiz, transcripts of the audio, and downloadable MP3 files.

Livemocha http://www.livemocha.com/ is an innovative site that enables you to practice conversing with text and audio chat tools. In addition to self-study lessons, you can receive one-on-one instruction from tutors and native speakers.

English as a second language podcast http://www.eslpod.com/website is especially useful for beginners. It provides the opportunity to listen to conversations and discussions spoken at a slow, easy-to-understand speed. This site highlights useful everyday phrases and expressions, explaining what they mean and how to use them.

Rachel's English http://www.rachelsenglish.com is a videoblog with a text script. It concentrates on useful language and how to pronounce it, and it has links to many exercises.

Audiovisual Material about Diseases and Procedures

MedlinePlus Interactive Health Tutorials http://www.nlm.nih.gov/medlineplus/tutorial.html, directed at patients, this page provides animated graphics explaining a procedure or condition in easy language.

http://www.nlm.nih.gov/medlineplus/videosandcooltools.html contains interactive tutorials and anatomy videos with transcripts to allow you listen and read simultaneously.

http://www.nlm.nih.gov/medlineplus/surgeryvideos.html, more directed toward healthcare professionals, this page provides links to prerecorded webcasts of surgical procedures. These are actual operations performed at medical centers in the United States since 2004. These one-hour-long videos are accompanied by PDF transcripts.

i on NIH http://ocplmedia.od.nih.gov/vodcast/20100101NIHvodcast_0024.mp4 is a 30 min long monthly vodcast summarizing important advances from medical research at the all 27 of the National Institutes of Health.

The Discovery Channel's Health Section http://health.discovery.com/ contains a wealth of video material as well as interesting interactive features.

VuMedi www.vumedi.com is a surgeon-only video sharing website where you can view, upload, and discuss surgical videos. The site is free but registration is required.

O.R. Live http://www.orlive.com contains surgical webcasts from the University of Maryland Medical Center that allow you to watch surgeons operate and listen to them explain the procedures.

Audiovisual Material to Improve Public Speaking Skills

http://www.americanrhetoric.com This website contains hundreds of famous speeches. Each speech has the transcript with it. You can hear the audio and sometimes see the video of the speeches, too.

Communication Steroids http://www.communicationsteroids.com has a series of texts and podcasts with many useful tips to help you improve your public speaking.

General Audiovisual Material

Television

60 minutes http://www.cbsnews.com/sections/60minutes/main3415.shtml provides short excerpts from CBS's investigative journalism program

CNN http://edition.cnn.com/video offers live video, podcasts, and CNN radio.

Where to watch TV online http://www.seabreezecomputers.com/tips/tv.htm provides a long list of full-length video-on-demand online television programs.

Radio

National Public Radio (NPR) http://www.npr.org, America's premier nonprofit, nongovernmental radio station's website provides noncommercial news, talk, and entertainment programming. They have a special "health & science" section and programs can be downloaded as podcasts.

BBC http://www.bbc.co.uk/radio, the world's largest broadcaster, provides public service broadcasting including downloadable news programs and documentaries.

Audiovisual Material Related with Major Medical Journals

Emerging Infectious Diseases www.cdc.gov/EID has a weekly podcast with the added bonus of accompanying transcripts.

The American College of Cardiology's Conversations with Experts http://www.cardiosource.org/Certified-Education/eLearning-and-Products/Conversations-with-Experts.aspx provides 20 min discussions on a single topic between leading experts in cardiology designed to facilitate peer-to-peer discussions on topics of importance to the cardiology community. You can watch these online with accompanying abstracts, speaker biographies, and transcripts or download the podcast.

Johns Hopkins PodMed http://www.hopkinsmedicine.org/mediaII/Podcasts.html, oriented toward patients, this podcast provides 5- to 7-minute dialogues about the week's medical news. Although there are no transcripts, an outline is available and there is a commentary about the podcast at http://hopkinspodblog.wordpress.com/

JAMA Audio Commentary http://jama.ama-assn.org/misc/audiocommentary.dtl consists of 10 min long summaries and comments on the week's issue of the journal.

Their *Author-in-the-room podcasts* http://jama.ama-assn.org/authorintheroom/authorindex.dtl provide in-depth interviews (about 1-hour long) with the authors of articles. Although no transcripts are available, written summaries are provided.

NEJM this week http://podcast.nejm.org/nejm_audio_summaries.xml provides audio summaries of the week's issue of the New England Journal of Medicine.

NEJM Audio Interviews http://podcast.nejm.org/nejm_audio_interview.xml

NEJM Videos in Clinical Medicine www.nejm.org/multimedia/videosinclinicalmedicine requires a subscription to the journal, but enables different versions to downloaded for windows media, palm computers, or MPEG-4 and accompanying PDF texts.

Nature http://www.nature.com/nature/videoarchive is an online video streaming archive featuring interviews with scientists behind the research and analysis from *Nature* editors about selected articles. *Nature* also has a YouTube channel http://www.youtube.com/NatureVideoChannel?gl=GB&hl=en-GB that enables you to easily upload and share their videos across the Internet through websites, mobile devices, blogs, and email.

http://www.nature.com/nature/podcast features a free audio show with highlighted content from the week's edition of Nature including interviews with the people behind the science, and in-depth commentary, and analysis from journalists covering science around the world. Best of all, transcripts are available for all podcasts in *Nature Medicine* www.nature.com/nm/podcast and *Nature Neuropod*

Neurology http://www.aan.com/rss/?event=feed&channel=1 has a weekly podcast discussing several highlighted articles in the current issue of the journal. It also features an interview with the author of an article in the current issue that summarizes the paper and discusses the main findings and clinical implications for neurologists. It concludes with the Lesson of the Week, a short segment on a topic such as a laboratory technique, statistical methods, or historical neurology. Podcast listeners can earn 0.5 AMA PRA Category 1 CME Credits by answering the multiple-choice questions in the online Podcast Quiz.

The Lancet www.thelancet.com/audio features free podcasts with author interviews and expert discussions on clinical practice, evidence, and campaigns. There are separate, additional audio archives for infectious diseases, neurology, and oncology.

The BMJ Group Podcasts http://podcasts.bmj.com includes different archives of audio files from their different member journals.

The Journal of Immunology's Immunocasts http://www.jimmunol.org/rss/jipodcast.dtl highlight papers regarded by reviewers and editors as the top 10% in the field. Transcripts are provided, but only subscribers have access to the material.

Podcasts from Medical Schools and Teaching Hospitals

UT Internal medicine residency podcasts http://www.utmemphis.libsyn.com enables you to listen in on the grand rounds, intern school, and noon conference, either online or by downloading to an MP3 player

The University of Indiana's School of Medicine produces *Sound Medicine* http://soundmedicine.iu.edu/, an hour-long show with interviews with medical experts about a wide range of current issues in medicine.

Vanderbilt University produces *ICU rounds* http://icurounds.com/, an archive of talks about different issues in critical care medicine. Unfortunately, there are no tapescripts or other supporting material.

Congress Webcasts

Webcasts of congress lectures not only offer an excellent way to keep up-to-date with new developments in medicine in science, but they also give you the opportunity to practice your English and hone your presentation techniques. Webcasts are often available at more than one site, as congress organizers and sponsors often post the material. It is worth the effort to look for sites that provide extras like printable PDF scripts or simultaneous subtitling. Although some organizations only give free access to members or attendees and others package the material as pay-for-view CME material, many provide free access to all. Here is a list of just a few of the congress webcasts revealed by a quick search on the internet:

American Association for Cancer Research www.aacr.org has audio, video, and slide presentations. This material is free and open to everyone.

American Society of Colon and Rectal Surgeons www.fascrs.org has only slides and audio, but is free and open to all.

American Society of Human Genetics www.ashg.org has material directed at audiences with different levels of expertise; it is open to everyone.

American Society for Reproductive Medicine www.asrm.org has material for patients and professionals as well as presentations from the congress. It is open to everyone.

European Association of Urologists www.uroweb.org is open to everyone.

European Society of Cardiology Congress Webcasts www.escardio.org/**congress**es/esc-2009 contains 250 presentations from 2009 alone, as well as many others from other congresses. It is open to everyone.

European Cystic Fibrosis Society www.ecfs.eu is open to everyone.

European Hematology Association www.ehaweb.org is open only to members or those who attended the congress.

Oncology Nursing Society www.ons.org is open to everyone.

International Preterm Labour Congress www.oblink.com has presentations from the annual meeting and the transcripts!

The Informed Scientist http://www.informed-scientist.org is a free online platform for scientific communications and presentations in the fields of medicine and biology. At the time this was written, the site had webcasts from the Fifth Congress of the European Crohn's and Colitis Organisation, from the Sixth Central European Gastroenterology Meeting, and from the Sixth International St. Gallen Oncology Conference, amongst others.

MIX
Papier aus verantwortungsvollen Quellen
Paper from responsible sources
FSC® C105338

If you have any concerns about our products,
you can contact us on
ProductSafety@springernature.com

In case Publisher is established outside the EU,
the EU authorized representative is:
Springer Nature Customer Service Center GmbH
Europaplatz 3, 69115 Heidelberg, Germany

Printed by Libri Plureos GmbH
in Hamburg, Germany